Regular Algebra
Algebra
AND
Finite
Machines

Regular Algebra
Algebra
AND
Finite
Machines

John H. Conway

John von Neumann Professor in
Applied and Computational Mathematics
Princeton University

DOVER PUBLICATIONS, INC.
Mineola, New York

Bibliographical Note

This Dover edition, first published in 2012, is an unabridged republication of the work originally published in 1971 by Chapman and Hall, Ltd., London.

Library of Congress Cataloging-in-Publication Data

Conway, John Horton.
　　Regular algebra and finite machines / John H. Conway.
　　　　p. cm.
　　This Dover edition, first published in 2012, is an unabridged republication of the work originally published in 1971 by Chapman and Hall, Ltd., London.
　　Includes index.
　　ISBN-13: 978-0-486-48583-6
　　ISBN-10: 0-486-48583-8
　　1. Sequential machine theory. 2. Algebra. I. Title.

QA267.5.S4C65 2012
511.3'5—dc23

2012003545

Manufactured in the United States by Courier Corporation
48583801
www.doverpublications.com

Contents

CONTENTS

Preface

This book has been growing for some time. In 1960 I was particularly impressed by the papers of Kleene and Moore in *Automata Studies* (ed. Shannon and McCarthy), and decided to examine the axiomatics of Kleene's algebra and to improve Moore's bounds for lengths of experiments. Neither problem made much progress for some years, but then Michael Guy and I found the method of Chapter 2 for the Moore problem, and some time later I developed the treatment of the axiomatic problem described in Chapter 13, and spoke briefly about it at the 1966 International Congress of Mathematicians in Moscow. The matrix proof of Kleene's theorem, and the differential calculus of events (Chapters 3 and 5), predate this. I have since learnt that the matrix formula was already known to P. J. Cleave in 1961, while J. Brzozowski has published several papers on the differential calculus in more recent years.

I should like to thank in particular Andrew Glass, whose lucid notes put some sense into a lecture course I gave on these topics in Cambridge in 1966, and from which large parts of Chapters 0–6 and 12–13 have been copied almost verbatim, and my former research student Donald Pilling, whose work is the basis of Chapters 7–11. The idea of stressing the operator calculus is Pilling's, and most of the results of Chapters 7–11 appear in some form or other in his PhD dissertation, even when this is not explicitly acknowledged in the text. In addition to his work on the operator calculus, Pilling seems to have given the first complete treatment of commutative regular algebra (Chapter 11).

vii

I should also like to thank Peter Rowat, whose earlier work helped to lay the foundations for some of the book, and Michael Guy for the contributions noted above.

Not many references to the literature appear in the text, largely because I am not as familiar with the literature as I should be. For the same reason, most of the notation of the book is somewhat unorthodox, but I have taken some trouble to make it self-consistent, and hope that at least some of it will be found generally useful (excluding of course the bizarre names for particular operators). I hope that more than one reader will find the time to attack the various conjectures made, particularly in Chapters 6, 9, 11, 13 and 14. Not all of these can really be as hard as I have found them!

Preliminaries to the Moore theory

The main aim of this book is a presentation of Kleene's algebraic theory of machines. To this end, we begin by formalizing the notion of machine. Then we present Moore's identification theory of machines, which logically seems to precede the Kleene theory.

Any actual machine is finite in the obvious sense that it has but a finite number of component parts, and so our theory will naturally be a theory of *finite* machines. Any actual machine is also to some extent unreliable, so that its behaviour is not always that which would be predicted from a knowledge of its design. However, we choose to ignore this unsavoury phenomenon, arguing that the deterministic theory is at once simpler and more interesting, and can in any case be made to include the probabilistic theory by the simple device of adding new input facilities accessible only to gremlins.

Any machine worth the name will want to communicate with its environment by means of suitable input and output facilities. If the machine is deterministic, its output signals at any time will depend only on the state of the machine at that time, and the action of the machine at any instant will depend only on its state at that instant and the nature of any input signals it may receive.

We there formalize the notion of *finite machine* (*sequential machine, Moore machine, finite automaton*) as follows:

A finite machine M consists of:
    a finite set $S = S(M)$ of objects called *states*
    a finite set $I = I(M)$ of objects called *inputs*

a finite set $O = O(M)$ of objects called *outputs*
a function $t: S \times I \to S$ called the *transition function*
a function $o: S \to O$ called the *output function*
a particular state ι called the *initial state.*

We customarily use the letters $\alpha, \beta, \gamma, \ldots$ for states, $a, b, c, \ldots$ for inputs, and $i, j, k, \ldots$ for outputs, and we abbreviate $t(\alpha, a)$ to α_a.

The appropriate mental picture is this. At any time the machine is in a given state α, continuously emitting output $o(\alpha)$, and it will remain in state α until it receives an input signal a, say, when it will instantly assume the state α_a and emit output $o(\alpha_a)$. When the machine is delivered from the manufacturers, it is in the initial state ι.

If we think of the machine as a black box with push-buttons for inputs and lights for outputs, then the buttons must be spring-loaded so as to pop out after each pressing, and mechanically interlocked so that only one can be operated at any instant. Again, the electrical circuitry operating the outputs must be so designed that exactly one lamp is lit at any time. Other types of machine are readily accommodated by various coding processes – thus the formalized version of a machine with two input buttons a and b which can be operated separately or simultaneously, and two lamps i and j which can be lit or unlit independently of each other, will not have two inputs and two outputs, but rather three inputs ($a, b, both$) and four outputs ($i, j, both, neither$).

Another popular formalization should perhaps be mentioned. A *Mealey machine* is defined just as a Moore machine, except that the output function is from $S \times I$ to O rather than simply from S to O. We think of such a machine as emitting its outputs at the instant of transition from one state to another, the output depending both on the previous state and the disturbing input. The theories of the two types are obviously equivalent, since any Mealey machine can be imitated by a Moore machine which has more states and 'remembers' the previous input, and a Moore machine can obviously be regarded as a Mealey machine in which the output happens to be independent of the input. We use the Moore version because it fits more simply into the Kleene theory.

We can describe a machine most conveniently by tables giving the values of the functions α_a and $o(\alpha)$, or graphically by means of its *state diagram*. The latter has a node for each state, marked with the name of that state and its output, and a directed branch marked a from node

α to node β whenever $\alpha_a = \beta$. The initial state is marked by an ingoing arrow ($\rightarrow$). When there are just two outputs 0 and 1 we call M a *binary machine*, and in this important case we can alternatively indicate the outputs by marking outgoing arrows ($\alpha\rightarrow$) from those states α with $o(\alpha) = 1$. These remarks should be made quite clear by the example presented in Fig. 0.1.

t	a	b	o
$\rightarrow\alpha$	β	γ	0
β	γ	γ	0
γ	γ	α	1

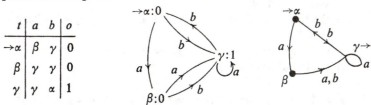

Fig. 0.1. A three-state machine and its state diagram.

The Moore theory considers experiments on a finite machine M, under the supposition that the experimenter is forbidden to examine the state of M by any means other than observing its output. The aim is to extract information about the structure or state of M by applying various carefully chosen inputs and studying the resulting output-sequences. The following terminology is useful.

A *word* is a finite (possibly empty) formal sequence of inputs. The *length* of the word is the length of this sequence, and the *empty word* 1 is the word of length zero. I^* denotes the set of all words in inputs of I. If α is a state and w is the word $ab \ldots k$, then α_w denotes the state $(\ldots((\alpha_a)_b)\ldots)_k$ obtained from α by applying the inputs $a, b, \ldots, k$ in order, and naturally $\alpha_1 = \alpha$. A state β is *accessible from* α only if there is a word w with $\alpha_w = \beta$, and *accessible* if it is accessible from ι.

It is obvious that all observable properties of the state α are properties of the function $f_\alpha : I^* \rightarrow O(M)$ defined by $f_\alpha(w) = o(\alpha_w)$. It is therefore natural to call α and β *indistinguishable* if and only if $f_\alpha = f_\beta$ – that is to say, if and only if $o(\alpha_w) = o(\beta_w)$ for every word w. Note that this definition can be meaningful even if α and β are states of distinct machines M and M' – all we need require is that M and M' have the same *console*; that is to say, the same sets I and O.

The engineer who wants a machine for some specific purpose will normally prefer the simplest machine that does the job. He will not usually approve of a multiplicity of parts with the same effect, and nor will he countenance the insertion of components with no function. These considerations lead naturally to the following notions.

The quotient by indistinguishability

For any machine M, we define a machine $M^{\div}$ with the same console by:

- (i) the states of $M^{\div}$ are the indistinguishability classes $(\bar{\alpha})$ of the states (α) of M,
- (ii) the transition function is defined by setting $\bar{\alpha}_a = \bar{\beta}$ whenever $\alpha_a = \beta$,
- (iii) the output function is defined by $o(\bar{\alpha}) = o(\alpha)$,
- (iv) the initial state is $\bar{\iota}$.

Theorem 1. $M^{\div}$ *is indeed a machine, and any state α of M is indistinguishable from the corresponding state $\bar{\alpha}$ of $M^{\div}$.*

Proof. $M^{\div}$ is well-defined since indistinguishable states have the same output, and if α and β are indistinguishable states so are $\bar{\alpha}$ and $\bar{\beta}$. The states α and $\bar{\alpha}$ are indistinguishable, since $o(\bar{\alpha}_w) = o(\overline{\alpha_w}) = o(\alpha_w)$ for every word w.

The truncation by inaccessibility

For any machine M, we define a machine M^{-} with the same console by:

- (i) the states of M^{-} are the accessible states of M,
- (ii) the transition and output functions of M^{-} are the restrictions of those of M to the states of M^{-},
- (iii) the initial state of M^{-} is ι.

Theorem 2. M^{-} *is indeed a machine, and every state α of M^{-} is indistinguishable from α regarded as a state of M.*

Proof. M^{-} is a machine, since α_a is accessible whenever α is. The two states α are indistinguishable, since α_w has the same meaning in M^{-} as in M.

The point is, of course, that any two states of M^{+} are distinguishable, and any state of M^{-} accessible. Figure 0.2 illustrates the two constructions.

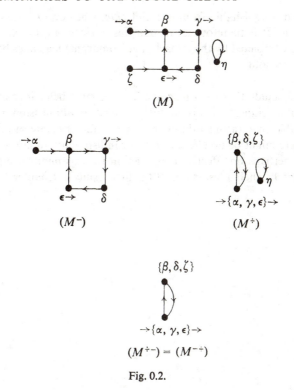

Fig. 0.2.

We say that M is *minimal* if every state of M is accessible and every two distinct states of M are distinguishable. We say that M and N have the same *behaviour* if and only if their initial states are indistinguishable. The notion of *isomorphism* between two machines with the same console is defined in the obvious way.

Theorem 3. $M^{\div-}$ *and* $M^{-\div}$ *are minimal machines with the same behaviour as* M. *Minimal machines have the same behaviour if and only if they are isomorphic.*

Proof. The first statement follows easily from our previous theorems. If now M and N are minimal machines with the same behaviour and initial states α and β, we define a function from the states of M to those of N by $f(\alpha_w) = \beta_w$ for each word w. Then f is *everywhere defined*, since each state of M is accessible; *well-defined*, since β_w is the unique state of N indistinguishable from α_w; and *onto*, since every state of N is accessible. It is therefore an isomorphism, since $o(\alpha_w) = o(\beta_w)$. The proof is completed by the remark that isomorphic machines have the same behaviour.

We conclude that each machine has the same behaviour as some minimal machine, which is uniquely defined to within isomorphism. A possible snag in this otherwise very satisfactory state of affairs is that there might be no effective test for indistinguishability of states, and so perhaps no effective construction for this minimal machine. Fortunately, such a test is one of the first results of Chapter 1.

CHAPTER ONE

Moore's theory
of experiments

The maintenance man in charge of a real machine M often has occasion to experiment with M. Sometimes he will be presented with M in some unaccustomed (and probably unknown) state and required to return it with a note specifying its new state. Occasionally he might suspect that some malfunction has transformed M into a new (and probably useless) machine M', and will need to devise an experiment to determine whether this has in fact occurred. In both cases his method will probably be to apply certain inputs to the machine (each input depending on the previous outputs) until the resulting sequence of outputs in some sense contains the information he requires. The outcome of the experiment will be a member of some set of answers, for instance the set {*yes, no, don't know*}.

Formally, we define an *experiment* on M as a function $e: O^* \to I \cup A$, where O^* is the set of all *output words* (i.e., finite formal sequences of outputs), and A is some set of *answers*, disjoint from I. We perform e on M as follows. At any stage we will already have applied a word $w = ab \ldots k$, and observed the *resulting* output word

$$r(\alpha, w) = o(\alpha), o(\alpha_a), o(\alpha_{ab}), \ldots, o(\alpha_{ab\ldots k}).$$

If $e(r(\alpha, w))$ is an input l, say, we extend w by applying l to the machine and observing the new output $o(\alpha_{ab\ldots kl})$, so proceeding to the next state. If not, $e(r(\alpha, w))$ is an answer called the *outcome* of e at α, and the experiment terminates.

We can perform the experiment at any state of any machine with the same console as M, and its outcome, if any, will probably depend on both machine and state. If in all intended performances it terminates in a bounded number of stages, we call it *finite*, and define its *length* as the greatest length of any of the corresponding input words w. (Note that the length of an input word w is one less than that of the resulting output word $r(\alpha, w)$, since the first output 'comes free'.) We shall mostly be concerned with finite experiments.

The aristocrat's experiment is simpler. He requires his man only to apply a certain word w to the machine and return with the resulting output word $r(\alpha, w)$, which should contain all the information he desires. In the formalized version $e(r(\alpha, v))$, if an input, will be determined already by the length of $r(\alpha, v)$ (or of v), being independent of the individual outputs. The aristocrat pays for his pleasures – the construction of a word w to elicit certain information is harder than the construction of a more general experiment, and might well be impossible. But the aristocrat's experiment is easy to specify – we speak of 'the experiment w'.

Definitions. An (n, m, p)-*machine* is one with at most n states, at most m inputs, and at most p outputs. An *exact* (n, m, p)-machine is one with exactly n states, exactly m inputs, and exactly p outputs, each of which is the output of some state (so that the function o is onto).

Theorem 1. *Two distinguishable states of an exact (n, m, p)-machine can be distinguished by some word of length at most $n - p$.*

Proof. We define $\alpha =_N \beta$ to mean that $o(\alpha_w) = o(\beta_w)$ for all input words w of length at most N, and $\alpha =_\infty \beta$ to mean that α and β are indistinguishable. These are equivalence relations: let $f(N)$ denote the number of equivalence classes of $=_N$. Then plainly

$$f(0) \leqslant f(1) \leqslant f(2) \leqslant \ldots \leqslant f(\infty) \leqslant n,$$

and so we can define N_0 as the least N with $f(N) = f(N + 1)$. But since

(i) $\alpha =_0 \beta$ if and only if $o(\alpha) = o(\beta)$
(ii) $\alpha =_{N+1} \beta$ if and only if $o(\alpha) = o(\beta)$ and $\alpha_a =_N \beta_a$ for all $a \in I$

the relation $=_{N_0+2}$ is obtained from $=_{N_0+1}$ just as is $=_{N_0+1}$ from $=_{N_0}$, and so $f(N_0) = f(N_0 + 1) = f(N_0 + 2) = \ldots = f(\infty)$, and we have

$$p = f(0) < f(1) < \ldots < f(N_0) = f(\infty) \leqslant n,$$

and so $N_0 + p \leqslant f(N_0) \leqslant n$, and any two distinguishable states are distinguishable by a word of length at most $N_0 \leqslant n - p$.

The argument also gives a practicable test for indistinguishability: compute in succession the relations $=_0, =_1, \ldots$ until two successive relations coincide, when both are $=_\infty$.

Theorem 2. *If S is a set of at most s states of an exact (n, m, p)-machine, and some two states of S are distinguishable, then there is a word of length at most $\max(0, n - p - s + 2)$ which distinguishes some two states of S.*

Proof. Let $N = n - p - s + 2$, and N_0 as in Theorem 1, so that we may suppose $N \leqslant N_0$. If $N \leqslant 0$, then $s \geqslant n - p + 2$, and since there can be at most $n - p + 1$ states with any given output (there are $p - 1$ states with other outputs) some two states of S must have distinct outputs and are already distinguished. So we may suppose $0 < N \leqslant N_0$, so that $f(N) \geqslant N + p$, and each $=_N$ class has at most $n - (N + p - 1) = s - 1$ members, so that some two states of S are in distinct $=_N$ classes.

Theorem 3. *Theorem 2 gives the best possible result as a function of n, m, p, s.*

Proof. Take a machine with states $\alpha_1, \ldots, \alpha_n$ and $(\alpha_i)_a = \alpha_{\min(i+1, n)}$, $o(\alpha_i) = \max(1, i - n + p)$, with S the set $\{\alpha_1, \ldots, \alpha_s\}$. In Fig. 1.1 we have $n = 7$, $m = 2$, $p = 3$, $s = 2$. Then to take any state of S to a state of output other than 1 we require a word of length at least $\max(0, n - p - s + 2)$.

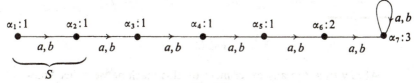

Fig. 1.1.

Theorem 4. *If we are given the structure of an exact (n, m, p)-machine whose states are all distinguishable, and told that it is initially in one of a set S of at most s states, then we can specify an experiment of length at most $(\bar{s} - 1)(n - p - \frac{1}{2}(\bar{s} - 2))$ after application of which the final state will be known, where $\bar{s} = \min(s, n - p + 2)$.*

Proof. If $s > \bar{s}$ some two states of S give distinct outputs, so that at least one can be eliminated before we start. So we suppose $s = \bar{s} \leqslant n - p + 2$. By Theorem 2 there is a word of length at most $n - p - \bar{s} + 2$ after which we can reduce the size of S. The new S has at most $\bar{s} - 1$ elements, and so there is a word of length at most $n - p - (\bar{s} - 1) + 2$ which distinguishes some two of its elements. Proceeding in this way we eventually obtain an experiment of length at most

$$((n - p) - (\bar{s} - 2)) + ((n - p) - (\bar{s} - 3)) + \ldots + (n - p)$$
$$= (\bar{s} - 1)((n - p) - \tfrac{1}{2}(\bar{s} - 2))$$

after which the final state is known.

Theorem 5. *The function of Theorem 4 is the best possible as a function of n, s, p, although it may be improved if m is taken into account.*

Proof. Take the machine M_n with states $\alpha_1, \ldots, \alpha_n$, output function $o(\alpha_i) = \max(1, i - n + p)$, and S the set $\{\alpha_1, \ldots, \alpha_s\}$, but with $n - p + 1$ inputs $a_k(1 \leqslant k \leqslant n - p + 1)$ such that a_k interchanges α_k and α_{k+1} but leaves all other states unchanged. (We could also add further inputs leaving all $\alpha_k(k \leqslant n - p + 1)$ unchanged.) If we define the *depth* of any state α_i as $n - p + 2 - i$, and the depth of any set of states as the sum of their depths, then the states of positive depth are just those with output 1. In Fig. 1.2 we illustrate the case $n = 7$, $p = 3$, $s = 2$, $m \geqslant 5$, with undrawn transitions taking states to themselves.

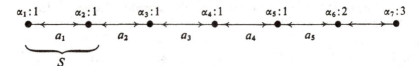

Fig. 1.2.

At any time t in any experiment on this machine we either know the state or know that it is in one of a set S_t of states of positive depth, and it is easy to see that $\text{depth}(S_{t+1}) \geqslant \text{depth}(S_t) - 1$. But to determine the state uniquely we must reduce the depth to $\text{depth}(\alpha_1) = n - p + 1$, and so the experiment must have length at least

$$\text{depth}(S) - \text{depth}(\alpha_1) = (s - 1)(n - p - \tfrac{1}{2}(s - 2)),$$

assuming – as we may – that all states of S have positive depth.

Note. Analysis of the argument shows that M_n is essentially the only machine which needs an experiment of the full length, and so we can improve the bound if $m \leqslant n - p$. But Moore gives a more complicated example which shows that the bound is within a constant factor of the true result even for $m = 2$.

As corollaries of these theorems we have:

Theorem 6. *Two distinguishable states of any (n, m, p)-machine can be distinguished by a word of length at most $n - 2$.*

Proof. We apply Theorem 1, noting that $p \geqslant 2$, since two states are distinguishable.

Theorem 7. *Two distinguishable states of different (n, m, p)-machines can be distinguished by a word of length at most $2n - 2$.*

Proof. We consider the *direct sum* of the two machines, whose state diagram is the union of disjoint state diagrams of the two machines, the initial state being unimportant. We then apply Theorem 6 to the $2n$ states of this disconnected machine to obtain the result.

Theorem 8. *The final state of an (n, m, p)-machine whose states are all distinguishable can be found by an experiment of length at most $\frac{1}{2}(n - 1)(n - 2)$.*

Proof. Again, $p \geqslant 2$, and we can put $s = n - p + 1$ in Theorem 4.

We have arranged the numbering so that our Theorems 6, 7, 8 and 9 correspond to Moore's, but Moore's Theorems 6, 7 and 8 have the numbers $n - 1$, $2n - 1$, $\frac{1}{2}n(n - 1)$. The discrepancy is due to a difference in conventions – Moore does not allow himself to look at the final output of the sequence resulting from his experiment. The two versions of Theorems 6 and 7 agree when this is taken into account, but Moore's curious convention has kept him from the true result for Theorem 8.

We now consider the problem of determining the structure of a machine when we are told only that it belongs to a set of possibilities all of which have the same console.

Theorem 9⁻. *If we are given a set S of at most s (n,m,p)-machines all of whose states (in different machines or the same machine) are distinguishable, then we can identify an unknown machine from S by an experiment of length at most $\frac{1}{2}n^2 s^2$.*

Theorem 9⁺. *With the same hypotheses, an experiment of length at most $n^2 s$.*

Proof of 9⁻. The direct sum of all the machines of S has at most ns states, and so there is an experiment of length at most $\frac{1}{2}(ns-1)(ns-2)$ which determines it final state. But the problem of identifying an unknown machine from S is the same as that of determining which summand contains an unknown state of the direct sum.

Proof of 9⁺. Suppose the direct sum were initially in one of the n states corresponding to the first machine of S, and perform an experiment of length, at most, n^2, reducing it to a known state under this supposition. Now suppose that it was initially in one of the $2n$ states of the first and second machines of S, so that now it is in one of at most $n+1$ known states, and perform another experiment of length at most n^2 reducing it to at most 1 state under this supposition, and so on. In this way we obtain an experiment of length at most $n^2 s$ which determines the final state of the direct sum.

Theorem 9. *The set $R_{n,m,p}$ of all (n,m,p)-machines with all states distinguishable and all states of any machine accessible from each other has an identifying experiment of length at most $n^{nm+2} p^n/n!$*

Proof. We first supplement each such machine to a machine with exactly n states by adding new inaccessible states in such a way that all states of the new machine are distinguishable. Then the number of machines in $R_{n,m,p}$ is at most $n^{nm} p^n/n!$, for the supplemented machine has transition function chosen from a list of n^{nm} possibilities, output function chosen from a list of p^n possibilities, and appears (to within isomorphism) $n!$ times in the product of these two lists, the initial state being unimportant. We apply Theorem 9⁺ with $s = n^{nm} p^n/n!$

Machines satisfying the hypotheses of Theorem 9 are said to be in *Moore reduced form*, and our result is Moore's Theorem 9. Some hypothesis is necessary if we are to distinguish between any two distinguishable machines of S, since Fig. 1.3 shows three machines A, B, C for which A can be distinguished from B or C (by applying a), C from A or B (by applying b), but any experiment distinguishing A from B necessarily confuses B and C (since its first input must be a).

No significant improvement on Moore's Theorem 9 seems to have been published hitherto. Feeling that any improvement must require some essentially new idea, M. J. T. Guy and I produced the techniques described in Chapter 2, which yield comparatively short experiments (indeed *words*) which identify unknown machines. But P. F. Rowat

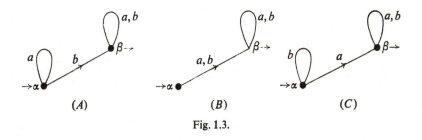

Fig. 1.3.

later produced the following simple argument which does almost as well, and has the merit of being in the same style as Moore's techniques.

Theorem 10. *Under the hypotheses of Theorems 9^- and 9^+, we can find such an experiment of length at most $m^{2n-2} . 4n \log s$.*

Proof. We consider the (at most s^2) pairs of distinct machines (M, N) from S, noting that each such pair is distinguished by one of the m^{2n-2} input words of length $2n - 2$, by Theorem 7. Apply such a word w_1 which distinguishes the greatest possible number of such pairs, and then we shall have at most $(1 - m^{2-2n})s^2$ pairs to be distinguished. It is plain that we can choose words w_1, w_2, . . ., w_k after application of all of which we have, at most, $(1 - m^{2-2n})^k s^2$ pairs to be distinguished.

We now observe that $\log(1 - m^{2-2n})^{-1} > m^{2-2n}$, and so if $k \geqslant 2m^{2n-2} \log s$ we shall have $(1 - m^{2-2n})^k s^2 < 1$, so that after application of the words w_1, . . ., w_k no pairs remain to be distinguished. The resulting experiment has length at most $(2n - 2)k$, and for suitable k establishes the Theorem.

Theorem 11. *In Theorem 9 we can replace the bound by* $m^{2n-1} \cdot 8n^2 \log n$.

Proof. We use the estimate $s \leqslant n^{nm} p^n \leqslant n^{mn+n} \leqslant n^{2mn}$ for the number of machines in $R_{n,m,p}$, so that $\log s \leqslant 2mn \log n$. We could improve the bound slightly by using better estimates.

CHAPTER TWO

Bombs and detonators

Consider two machines M and N with the same console. We should like to distinguish M from N if possible, and otherwise to do the best we can. Now, we cannot hope to distinguish M from N if our experiment takes them to indistinguishable states before taking them to states with distinct outputs, and simple examples show that we cannot hope to prevent this even when M and N are distinguishable. So we shall define an *identifier experiment* for a set S of machines as an experiment which, for any machines M, N of S, *either* distinguishes M from N *or* takes them to indistinguishable states. We ask which *words* constitute identifier experiments for S?

Definition. The *discriminant bomb* $M*N$ of M and N is the following machine:

(i) $M*N$ has a state α_{ij} of output 1 for each pair of distinguishable states α_i of M, α_j of N with the same output, and an *exploded* state α_* with $o(\alpha_*) = 0$.

(ii) If $(\alpha_i)_a = \alpha_I$ in M, and $(\alpha_j)_a = \alpha_J$ in N, where α_{ij} and α_{IJ} are both states of $M*N$, then $(\alpha_{ij})_a = \alpha_{IJ}$ in $M*N$.

(iii) Inputs in all other circumstances cause transition to α_*.

Theorem 1. *An experiment is an identifier for $\{M,N\}$ if and only if it always* explodes *$M*N$ (i.e., takes it from any state to the exploded state.)*

Proof. The experiment explodes $M*N$ if and only if it takes M and N to states with distinct outputs *or* to indistinguishable states.

Corollary. *An experiment is an identifier for a set S of machines if and only if it explodes every bomb $M*N$ $(M, N \in S)$.*

Theorem 2. *If M, N are (n, m, p)-machines and α is any state of $M*N$ there is a word w of length at most $2n - 2$ for which $\alpha_w = \alpha_*$.*

Proof. From Theorem 7 of Chapter 1.

So we consider the problem of *detonating* a given set of bombs.

Definition. An (n, m, h)-*bomb* is an $(n + 1, m, 2)$-machine such that:

   (i) one state α_* has output 0 – all others have output 1;
   (ii) all inputs fix α_*;
   (iii) For every state α there is a word w of length at most h with $\alpha_w = \alpha_*$.

The *height* of a set of states, or of a set of bombs, is the length of the shortest w for which $\alpha_w = \alpha_*$ for every α of the set, or for every α of every bomb of the set. A word w with this property is called a *detonator* for the set.

Theorem 3. *A set S of at most s (n, m, h)-bombs has height, at most, $2m^h \log 2ns$.*

Proof. Let $B \in S$. A *transition matrix* of B is an $(n + 1) \times (n + 1)$ matrix M with a row and column for each state $(\alpha_1, \ldots, \alpha_n, \alpha_*)$ of B and with M_{ij} equal to the number of inputs taking α_i to α_j.

We first estimate the number of l-words (words of length l) which are not detonators for B. Now, $(M^l)_{ij}$ is the number of l-words taking α_i to α_j, and so the number of non-detonator l-words is at most $v' M^l v$, where v is the column vector with $v_i = o(\alpha_i)$, for this is the sum over all $i, j \neq_*$ of the number of l-words taking α_i to α_j. To estimate $v' M^l v$ we use a trick whose origin and motivation will be explained later.

Let w be the column vector with $w_i = 1 + \varphi + \varphi^2 + \ldots + \varphi^{h_i-1}$, $w_* = h_* = 0$, where h_i is the height of $\{\alpha_i\}$ and φ the unique positive root of the equation $1 + \varphi + \ldots + \varphi^{h-1} = ((m-1)\varphi)^{-1}$.

Lemma. *We have $Mw \leqslant \varphi^{-1} w$, coordinatewise.*

Proof. The ith component $\sum_j M_{ij} w_j$ of Mw is the sum, with repetitions counted, of those w_j for which α_j can be obtained from α_i by a single input. If $h_i = 0$, then each $w_j = 0$, but also $w_i = 0$, and the result follows. Otherwise, some α_j has height $h_i - 1$, and so

$$\sum_j M_{ij} w_j \leqslant \frac{m-1}{(m-1)\varphi} + 1 + \varphi + \ldots + \varphi^{h_i-2}$$

(since the remaining $m - 1$ values of w_j are at most $((m-1)\varphi)^{-1}$), and the right-hand side of this is just $\varphi^{-1} + 1 + \ldots + \varphi^{h_i-2} = \varphi^{-1} w_i$.

Now, to estimate $v' M^l v$ we observe that $v \leqslant w \leqslant v/(m-1)\varphi$ componentwise, and so

$$v' M^l v \leqslant v' M^l w \leqslant v' \varphi^{-l} w \leqslant v' \varphi^{-l} v/(m-1)\varphi = n/(m-1)\varphi^{l+1},$$

since $v'v = n$. So the number of non-detonator l-words is at most $n/(m-1)\varphi^{l+1}$.

We now estimate φ. We have $1 + \varphi + \ldots + \varphi^{h-1} = 1/(m-1)\varphi$, and so plainly $(m-1)\varphi \leqslant 1$. Multiplying the first equation by $1 - \varphi$ we obtain $1 - \varphi^h = (1 - \varphi)/(m-1)\varphi$, which rearranges to

$$m\varphi = \left(1 - \left(\frac{m-1}{m}\right)\varphi^h\right)^{-1}$$

so that $m\varphi \geqslant 1$. From these we can deduce

$$\log m\varphi \geqslant (m-1)\varphi^h/m \geqslant (m-1)/m^{h+1}.$$

Now, for any bomb of S we have shown that there are at most $n/(m-1)\varphi^{l+1}$ l-words not detonators for that bomb, and it follows that there are at most $ns/(m-1)\varphi^{l+1}$ l-words not detonators for S.

Defining l_0 by the equation $m^{l_0} = ns/(m-1)\varphi^{l_0+1}$, we see that if $l > l_0$ there must be some l-word which *is* a detonator for S. But

$$l_0 + 1 = \log \frac{mns}{m-1} \Big/ \log m\varphi \leqslant m^{h+1} \log \frac{mns}{m-1} \Big/ (m-1)$$

by the above inequalities, and this is exceeded by the stated bound since $m/(m-1) \leqslant 2$, the case $m = 1$ being trivial. This concludes the proof of Theorem 3.

The idea behind the proof may be clarified as follows. We wish to estimate $v'M^lv$, for all matrices M of a certain type. Now, it is well known that the 'size' of M^l is roughly $|\lambda|^l$, where λ is the largest eigenvalue of the matrix M. We therefore seek the worst value of λ over all bombs B of S. Intuitively, it seems likely that this corresponds

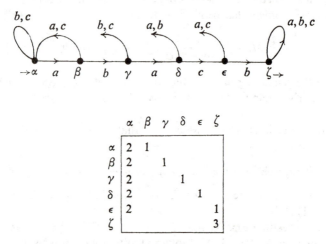

Fig. 2.1. A combination-lock bomb and its transition matrix.

to the most difficult bomb to explode, which we should expect to be the type of *combination-lock bomb* shown in Fig. 2.1. This has a sequence of states $\alpha_1, \ldots, \alpha_h, \alpha_*$, with just one input taking any state to the next in sequence, all other states either returning us to α_1 or being from α_* to itself. In Fig. 2.1 we give also the transition matrix, whose positive eigenvalue is φ^{-1} with corresponding eigenvector the w defined in the proof of our theorem.

Hence, for combination-lock bombs the inequality of the lemma becomes an equality, and our estimates for the number of non-detonators are correct to within small factors. If S contains even one combination lock bomb the estimate for the number of non-detonators for S is correct to within a small multiple of s, and since s finally appears only within the logarithm, Theorem 3 is best possible to within 'trivial' factors if regarded as an estimate of the l_0 such that *most l_0-words are detonators.*

Theorem 4. *If S has at most s (n,m,p)-machines, there is an identifier word for S of length at most $2m^{2n-2}\log 2n^2 s^2$.*

Proof. Put $(n,m,h,s) = (n^2, m, 2n-2, s^2)$ in Theorem 3.

Theorem 5. *The set $S = S_{n,m,p}$ of all (n,m,p)-machines has an identifier word of length at most $m^{2n}.2n\log 2n$.*

Proof. Let S^+ be the set of all (n,m,p)-machines with states $\alpha_1, \ldots, \alpha_n$ whose outputs form a consecutive sequence of integers starting with 1, $o(\alpha_i)$ being a monotonic increasing function of i, and let e be an

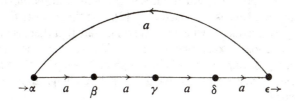

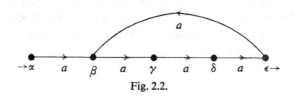

Fig. 2.2.

identifier for S^+. For $M \in S$, define M^+ as the machine obtained from M by renumbering the states and outputs of M in the order of their appearance during the application of e, states and outputs not appearing being numbered in any way which ensures $M^+ \in S^+$. If e fails to distinguish M from N, it will fail to distinguish some M^+ from some N^+, so take these to indistinguishable states, and so take M and N to indistinguishable states. It follows that e is an identifier for S.

Now, there are at most 2^{n-1} output functions for S^+, since each is specified by its difference function, a sequence of $n-1$, 0s and 1s. So there are $n^{nm}2^{n-1}$ machines in S^+. Applying Theorem 4 we obtain an identifier for S^+ of length at most $2m^{2n-2}\log(2^{n-1}n^{2nm+2})$, and this is exceeded by the stated bound provided $1 \leqslant 2^{m^2n-2n+1}n^{m^2n-2mn-2}$, which is always true, since if $m > 2$ both indices are positive, while for $m = 2$ it reduces to $n^2 \leqslant 2^{2n+1}$.

Like Theorem 3, Theorems 4 and 5 equally estimate the l_0 such that *most* l_0-words are identifiers, and again are essentially best possible in this respect. This follows from the fact that there are two n-state machines whose discriminant bomb is a combination-lock bomb of height $2n - 2$ (see Fig. 2.2 in which undrawn transitions are to the leftmost state).

We are left with the conclusion that a fairly certain way of identifying an (n,m,p)-machine is to sit at its keyboard and play a random sequence of inputs of length about $m^{2n}.2n\log 2n$, while bombs explode in the distance (one is reminded of 1812). The theory does not tell us how to pick a sequence which will work, merely that there is one. But to specify Moore's experiment in full would take about the same time as searching through all words of length $m^{2n}.2n\log 2n$, since his is a complicated experiment involving much branching, and ours is a single word.

We note that Moore gives examples of (n,m,p)-machines in Moore reduced form, showing that the best possible result is at least m^{n-1} (his *combination locks*, such as the left machine of Fig. 2.2). So, neglecting trivial factors, we can say that the true result is somewhere between m^n and m^{2n}. We shall illuminate this by a discussion of small machines at the end of the chapter.

The explorer problem

Moore remarks that a good reliability test for a machine would be an experiment causing it to undergo every possible transition of its state diagram at least once. If the machine is not in Moore reduced form this might not be possible, and so we call the word w an *explorer* for M if when w is applied to M, M undergoes every transition which is accessible from the state in which w leaves M.

Theorem 6. *A set S of at most s (n,m,p)-machines has an explorer of length at most $2m^n \log 2mn^2 s$.*

Proof. For any $M \in S$ and any transition of M we construct a bomb by replacing the corresponding branch of M's state diagram by a new branch to a new exploded state and then dropping states from which the new state is inaccessible. The number of bombs so obtained is at most nms, and all have height at most n, so we can apply Theorem 3 with mns for s and n for h.

Theorem 7. $S_{n,m,p}$ *has an explorer of length at most* $m^{n+1} . 2n \log 2n$.

Proof. We need consider only machines with inputs $a_1, \ldots, a_m$ and states $\alpha_1, \ldots, \alpha_n, \alpha_*$ for which $(\alpha_i)_a = \alpha_*$ if and only if $i = {}_*$ or $i = n$, $a = a_m$. The number of such machines is n^{nm-1}, since initial state and outputs are irrelevant. Putting $h = n$, $s = n^{nm-1}$ in Theorem 3 we get an explorer of length at most $2m^n \log 2n^{nm}$, and this is exceeded by the stated bound.

Some remarks on small machines

We have nowhere tried to find an experiment determining the initial state of a machine M, supposing the remaining structure of M is known. We explain this by reference to Fig. 2.3, a particular four-state machine. This machine is in Moore reduced form, but any experiment beginning with a confuses α and β, and any experiment beginning with b confuses β with γ. This state of affairs is Moore's *Uncertainty Principle*. By Theorem 6 of Chapter 1 it cannot happen with fewer than four states.

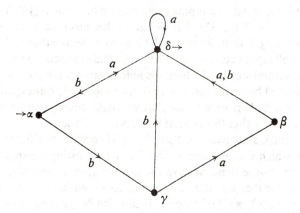

Fig. 2.3. Moore's Uncertainty Principle.

We define $f(m,n)$ as the minimal length of an identifier experiment for the set of all m input n state machines. We should like to know the best possible form of the index $g(n)$ in a result of the form $f(m,n) \leqslant f(n) \cdot m^{g(n)}$, and we expect a result roughly between n and $2n$. The following results were found in an attempt to clarify the situation by studying the small cases.

Theorem 8. $f(m,2) = 3m - 1$. $f(m,3) \leqslant 18m^2$.

Proof. For the first result, we observe that any state of a two-state machine may be named (to within indistinguishability) by its output. We now perform an experiment at each stage of which we move to a state from which there is a transition of unknown effect, then apply that transition and observe its effect. There are $2m$ stages, in the first $m + 1$ of which we need apply only one input per stage (there being

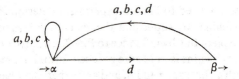

Fig. 2.4. The Devil's machine.

an unknown transition from the state we are in), and the last $m - 1$ stages take at most two inputs each, and we have $(m + 1) + 2(m - 1) = 3m - 1$. But if the Devil is designing his machine while we are experimenting with it, he will make it a machine like that of Fig. 2.4 in which all inputs except one take the first state to itself, and all inputs take the second state to the first, the unique input taking the first state to the second being the last one to occur as we apply our experiment, whatever it is. It is easy to see that we shall require at least $3m - 1$ inputs to verify that this is what the Devil has done.

For a three-state machine there is, by Theorem 6 of Chapter 1, an input x which distinguishes between any two distinguishable states, and if we choose these to have the same output, if possible, then it is easy to see that any state α is named to within indistinguishability by the pair $o(\alpha)$, $o(\alpha_x)$ of outputs. It can also be seen that if we apply the sequence x^3 to the machine we can determine the names of the states just before this application *and* just after it. We call x a *determining input*.

Now once we know a determining input x, we can 'buy' the state diagram at a cost of at most six inputs per transition, as follows. At any stage, we proceed, taking at most two inputs, to a state from which there is a transition of unknown effect; we then apply one input causing this transition; and follow up with the three inputs x^3 which determine the effect and also return us to a known state. Since there are $3m$ transitions in the state diagram this takes us $18m$ moves at most. If we do not know a determining input, we perform m such experiments, making each possible supposition in turn, since it is easy to show that we shall recognize the determining input when we have found it.

The results of Theorem 8, Moore's combination lock examples, and several other results, suggest that the true result has the form $f(m,n) \leqslant f(n)m^{n-1}$. But since the experiments in Theorem 8 relied heavily on the fact that the Uncertainty Principle fails for small machines, the evidence is not very conclusive.

We end with some even more special results. It is possible to show by hand that the sequence $ab^2 ab^3 aba^2 ba^3$ is an explorer for all two-input, three-state machines, and Guy has shown by a computer search that it is the unique such sequence of length 15, apart from the obvious possibility of interchanging a and b. Guy has also shown that the minimal length of an identifier word for the class of two-input, three-state machines is 26, and has found all the minimal identifier words, namely

aabbbbababbabbaaaababaabaa, *abbabbbbaabaaaababbbabaaab,*
aabbbbabbababbaaaababaabaa, *abbabbbbabaaababbbaabaaaab,*
aabbabbbbababbaaaababaabaa, *abbabbbbababbaabaaaababaab,*

and the words obtained by interchanging a and b in these. For three-input, two-state machines the minimal explorer length is 13, and there are just 29 minimal sequences, if we neglect the possibility of permuting the inputs. For two-state, two-input machines there is a unique minimal explorer with the same understanding, namely $a^2 bab^2$. These results show that identifier words must be longer than identifier experiments, since for a two-state machine there is no distinction between identifier words and explorer words.

CHAPTER THREE

Kleene's theory of regular events and expressions

Let M be a machine with initial state ι, and let i be any output of M. Define $E(i) = \{w \mid o(\iota_w) = i\}$, the set of all words taking M from its initial state to states of output i. Then the behaviour of M is characterized by the collection of sets $E(i)$ ($i \in O(M)$). Following Kleene, we ask what properties a collection of sets $E(i)$ must have if they are to be the $E(i)$ so derived from a finite machine. A set E of words is called *representable* if there is a finite machine M and an output i of M with $E(i) = E$.

Definitions. We recall that a *word* is a finite (possibly empty) formal sequence of inputs, and we now define an *event* as an arbitrary set of words. We do not distinguish between any input and the corresponding word of length 1, nor between any word and the corresponding event of cardinal 1, but we do distinguish between the *empty event* 0, which is the empty set of words, and the *unit event* 1, which is the set whose only member is the empty word. We call w an *occurrence* of the event E if $w \in E$.

This rather curious terminology is explained as follows. Kleene created his theory to describe the relation between an animal organism and it environment. The organism can detect an event taking place in the environment only by observing some special property of the input stimuli it receives, and so for mathematical purposes we can identify the event with the collection of all sequences of input stimuli which would correspond to occurrences of that event.

We shall now define various algebraic operations on events. The *sum* $E + F$ of two events is their set-theoretic union $E \cup F$. Their *product* EF is the set $\{ef \mid e \in E, f \in F\}$ of all words obtained by juxtaposing a word of E with a word of F. More generally, we can define arbitrary sums by writing

$$\sum_t^T E_t = \bigcup_t^T E_t,$$

the set-theoretic union of the sets E_t ($t \in T$), and then we can define the *star*, or *asterate*, of E by $E^* = 1 + E + E^2 + \ldots$, where the powers E^n are defined inductively by $E^0 = 1$, $E^{n+1} = E^n.E$. The nth partial sum of this expression is the nth *concatenated power* $E^{<n} = 1 + E + \ldots + E^{n-1}$, which can be inductively defined by $E^{<0} = 0$, $E^{<(n+1)} = E^{<n} + E^n$. We might also use $E^{\leqslant n}$ for $\sum E^i$ ($i \leqslant n$), $E^{>n}$ for $\sum E^i$ ($i > n$), etc.

The three operations $+$, $.$, $*$ are called the *regular operations*. Note that the infinitary sum operation $\sum$ is *not* considered as a regular operation, so that $*$ is in no sense a compound of the $+$ and $.$ operations. It is convenient to regard the constant event 0 as a nullary regular operation (i.e., an operation of scope zero), and we shall usually do so.

These operations satisfy the formal laws, called the *classical axioms*:

C1	$A + 0 = A$	C8	$A(B + C) = AB + AC$
C2	$A + B = B + A$	C9	$(B + C)A = BA + CA$
C3	$(A + B) + C = A + (B + C)$	C10	$(AB)C = A(BC)$
C4	$A.0 = 0$	C11	$(A + B)^* = (A^* B)^* A^*$ (sumstar)
C5	$0.A = 0$	C12	$(AB)^* = 1 + A(BA)^* B$ (productstar)
C6	$A.1 = A$	C13	$(A^*)^* = A^*$ (starstar)
C7	$1.A = A$	C14.n	$A^* = A^{n*} A^{<n}$ ($n > 0$) (powerstar)

Of these, only C11–14 require some explanation. We observe that $(A + B)^* = 1 + A + B + AA + AB + BA + BB + AAA + \ldots$, the sum of all products of A's and B's. On the other hand, the typical term of $(A^* B)^* A^*$ is $(A^i B)(A^j B) \ldots (A^m B)A^n$, the general product of A's and B's partitioned by occurrences of B. In a similar way we have

$$(AB)^* = 1 + AB + A(BA)B + A(BABA)B + \ldots = 1 + A(BA)^* B,$$

proving C12, while C13 is obvious, and C14.n asserts that every power A^i can be written in the form $(A^n)^j.A^k$, where $0 \leqslant k < n$. Note the abbreviation A^{n*} for $(A^n)^*$ – more generally, we treat the superscripts

*, n, $<n$ exactly as ordinary indices, so that for example we could further abbreviate C14 to $A^* = A^{n*+<n}$.

We call an event *regular* if and only if it can be obtained from the events 0, 1 and the inputs by repeated applications of the regular operations $+$, $.$, $*$, or, as we shall say, is a *regular function* of the inputs. For example, the event $E = (a^*b + bb^*a)^*$ is regular. It is the set of all words which can be written as a finite (possible empty) product of words of the form a^nb or bb^na ($n \geqslant 0$). This event E is also representable, since $E = E(1)$ for the machine of Fig. 3.1.

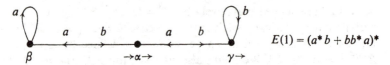

$$E(1) = (a^*b + bb^*a)^*$$

Fig. 3.1. An example of Kleene's main theorem.

This illustrates *Kleene's main theorem*:

Theorem 1. *An event E is representable if and only if it is regular.*

Proof. The proof relates the theorem to the theory of matrices whose entries are events. It will contain the proofs of several other theorems. In any machine M we define the event $E(\alpha, \beta)$ as the set $\{w | \alpha_w = \beta\}$ of all words taking the state α to the state β. We also define $E_l(\alpha, \beta) = \{w | \alpha_w = \beta$ *and w has length* $l\}$. The *event transition matrix* of M is the matrix defined by $M_{ij} = E_1(\alpha_i, \alpha_j)$, the set of all *inputs* taking α_i to α_j. We use the same letter for M and its transition matrix, which is an $n \times n$ matrix whose entries are events. The *input row vector L* of M is defined by $L_i = 1$ if α_i is the initial state, $L_i = 0$ otherwise. The *output column vector N* is defined by $N_j = o(\alpha_j)$. Its entries are outputs, but in the important case when M is a *binary* machine (with outputs 0 and 1 only) we identify the outputs 0 and 1 with the events 0 and 1.

It is easy to see that $E_l(\alpha_i, \alpha_j) = (M^l)_{ij}$, so that $E(\alpha_i, \alpha_j) = (M^*)_{ij}$, where M^* denotes the matrix $1 + M + M^2 + \dots$, the infinite sum being taken termwise, 1 denoting the usual $n \times n$ identity matrix, and the matrix products being defined in the usual way. So we must prove that the entries in M^* are regular events.

Given these preliminaries, it seems natural to embed our problem in a larger one. If we have any system of objects with operations

$\sum$, ., * satisfying the obvious laws, the $n \times n$ matrices over this system will satisfy (we hope) the same laws. If we have a subsystem closed under the regular operations $+$, ., * (where $+$ denotes the restriction of $\sum$ to two terms), can we assert that the $n \times n$ matrices over the subsystem will again be closed under $+$, ., *?

We define a *Standard Kleene Algebra* (**S**-*algebra*) to be a set S with three operations $\sum$, ., * defined on it (the **S**-*operations*), and particular elements 0 and 1 such that:

$$S1 \quad \overset{\emptyset}{\underset{t}{\sum}} E_t = 0 \ (\emptyset \text{ the empty set}) \qquad S4 \quad (EF)G = E(FG)$$

$$S2 \quad \overset{S}{\underset{s}{\sum}} \overset{T_s}{\underset{t}{\sum}} E_t = \overset{U}{\underset{t}{\sum}} E_t \left(U = \overset{S}{\underset{s}{\bigcup}} T_s \right) \qquad S5 \quad \overset{S}{\underset{s}{\sum}} E_s . \overset{T}{\underset{t}{\sum}} F_t = \overset{S \times T}{\underset{(s,t)}{\sum}} E_s . F_t$$

$$S3 \quad E.1 = 1.E = E \qquad S6 \quad E^* = \overset{\omega}{\underset{n}{\sum}} E_n \ (\omega = \{0, 1, 2, \ldots\}).$$

In $S5$, $S \times T$ denotes the cartesian product set of S and T. Of course, it is to be understood that $\sum_t^T E_t$ is defined for all index sets T, and denotes the sum of all the E_t $(t \in T)$. We note that $S6$ serves as a definition of E^*, and so we can (and often will) define an **S**-algebra simply by its $\sum$ and . operations. We define $E_0 + E_1$ as $\sum_t^{\{0,1\}} E_t$.

Theorem 2. *If we define $E \leqslant F$ to mean $E + F = F$, then $\leqslant$ is a partial order relation – indeed, a complete Boolean lattice relation – on any* **S**-*algebra.*

Proof. It is easy to verify the transitivity axioms, and if $E \leqslant F \leqslant E$ we have $E = E + F = F + E = F$. The least $X \geqslant E_t$ for all $t \in T$ is $\sum_t^T E_t$, since if $X \geqslant E_t$ for all $t \in T$ we have $X = \sum X = \sum (E_t + X) = \sum E_t + X$, all sums being over $t \in T$.

Theorem 3. *In any* **S**-*algebra, E^*G is the least F satisfying $F = G + EF$.*

Proof. E^*G is certainly such an F. On the other hand, any such F satisfies $F \geqslant EF$, and so $F \geqslant E^n F$ for each $n \geqslant 0$, whence $F \geqslant E^n G$, since $F \geqslant G$, and adding these inequalities for all n we deduce $F \geqslant E^*G$.

Theorem 4. *If $M = \begin{pmatrix} A & B \\ C & D \end{pmatrix}$ is a matrix over an* **S**-*algebra, partitioned* *in such a way that A and D are square, then*

$$M1 \qquad M^* = \begin{bmatrix} (A + BD^* C)^* & A^* B(D + CA^* B)^* \\ D^* C(A + BD^* C)^* & (D + CA^* B)^* \end{bmatrix}.$$

Proof. Let

$$M^* = \begin{pmatrix} E & F \\ G & H \end{pmatrix},$$

so that

$$\begin{pmatrix} E & F \\ G & H \end{pmatrix} = \begin{pmatrix} 1 & 0 \\ 0 & 1 \end{pmatrix} + \begin{pmatrix} A & B \\ C & D \end{pmatrix}\begin{pmatrix} E & F \\ G & H \end{pmatrix},$$

whence

$$E = 1 + AE + BG \qquad F = AF + BH$$
$$G = CE + DG \qquad H = 1 + CF + DH \qquad (*)$$

From two of these we deduce, using Theorem 3:

$$G \geqslant D^* CE \qquad \text{and} \quad F \geqslant A^* BH, \quad \text{and so}$$
$$E \geqslant 1 + AE + BD^* CE, \qquad H \geqslant 1 + CA^* BH + DH, \quad \text{whence}$$
$$E \geqslant (A + BD^* C)^* \qquad H \geqslant (D + CA^* B)^*, \quad \text{and so}$$
$$G \geqslant D^* C(A + BD^* C)^* \qquad F \geqslant A^* B(D + CA^* B)^*.$$

On the other hand, these values satisfy equations (*), and so define a matrix N satisfying $N = 1 + NM$, whence by Theorem 3, $N = M^*$.

The proofs given after the classical axioms show that these axioms are valid in any **S**-algebra, and also for matrices of conformable shapes over any **S**-algebra. Using them, we can transform the formula $M1$ into other forms, for instance

$$\begin{bmatrix} A & B \\ C & D \end{bmatrix}^* = \begin{bmatrix} (A^* BD^* C)^* A^* & (A^* BD^* C)^* A^* BD^* \\ (D^* CA^* B)^* D^* CA^* & (D^* CA^* B)^* D^* \end{bmatrix},$$

a form free of addition.

Theorem 5. *The entries in the star of any matrix M can be expressed as regular functions of the entries of M.*

Proof. The theorem is trivial for 1×1 matrices, and Theorem 4 provides the induction step from $a \times a$ and $b \times b$ to $(a + b)(a + b)$.

For example let

$$M = \left[\begin{array}{c|cc} 0 & a & b \\ \hline b & a & 0 \\ a & 0 & b \end{array}\right] = \left[\begin{array}{c|c} A & B \\ \hline C & D \end{array}\right],$$

say. Then

$$D^* = \begin{pmatrix} a & 0 \\ 0 & b \end{pmatrix}^* = \begin{pmatrix} E & 0 \\ 0 & F \end{pmatrix},$$

say, where $E = (a + 0b^*0)^* = a^*$, $F = b^*$, similarly. So

$$A + BD^*C = (a \quad b)\begin{pmatrix} a^* & 0 \\ 0 & b^* \end{pmatrix}\begin{pmatrix} b \\ a \end{pmatrix} = (aa^* \quad bb^*)\begin{pmatrix} b \\ a \end{pmatrix} = aa^*b + bb^*a,$$

or K, say. The formula now shows that

$$M^* = \begin{bmatrix} K^* & K^*aa^* & K^*bb^* \\ \cdots & \cdots & \cdots \\ \cdots & \cdots & \cdots \end{bmatrix}.$$

Since M was the event transition matrix of the machine of Fig. 3.1 this verifies Kleene's main theorem for that machine, and indeed we are ready for the proof of the first half of the theorem. This is because any representable event is $E(1)$ for some binary machine M, and so has the form LM^*N in terms of the transition matrix and initial and output vectors of M, and LM^*N is a regular event by Theorem 5. For the example of Fig. 3.1, $LM^*N = (aa^*b + bb^*a)^*bb^*$.

For the converse half of the Theorem, we observe first that an event has the form $E(1)$ for some machine if and only if $E = LM^*N$, where

(i) L is a constant row vector *with just one* 1.

(ii) M is a linear matrix *with each input occurring just once in each row.*

(iii) N is a constant column vector.

(An event is *constant* if it is 0 or 1, *linear* if it is a sum of inputs, and these adjectives apply to a matrix if and only if they apply to each of its entries.)

We can avoid the italicized restrictions by considering a generalized kind of machine. A linear mechanism is a set of objects α_i called *nodes*, together with three functions:

(i) the *initial function*, assigning to each α_i a constant event L_i,

(ii) the *transition function*, assigning to each pair α_i, α_j a linear event M_{ij},

(iii) the *output function*, assigning to each α_i a constant event N_i.

We specify a linear mechanism by the appropriate row vector L, square matrix M, and column vector N, or by a state diagram such as that of Fig. 3.2, which is like that for a binary machine but with certain restrictions lifted.

In the interpretation, a linear mechanism is thought of as a special kind of machine. At any time, a subset of the nodes, called the *state*, will be *active*, initially just those nodes α_i with $L_i = 1$. The node α_j will be active immediately after the application of some input a if and only if there was some α_i active immediately before, and $a \in M_{ij}$. The mechanism gives the output 1 if and only if some active α_j has $N_j = 1$.

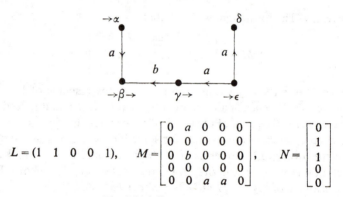

$$L = (1 \quad 1 \quad 0 \quad 0 \quad 1), \quad M = \begin{bmatrix} 0 & a & 0 & 0 & 0 \\ 0 & 0 & 0 & 0 & 0 \\ 0 & b & 0 & 0 & 0 \\ 0 & 0 & 0 & 0 & 0 \\ 0 & 0 & a & a & 0 \end{bmatrix}, \quad N = \begin{bmatrix} 0 \\ 1 \\ 1 \\ 0 \\ 0 \end{bmatrix}$$

Fig. 3.2. A linear mechanism.

The event represented by the mechanism is the event $LM*N$, which consist of all words w which take the initial state to some state of output 1. Formally, we have

Theorem 6. *E is represented by a linear mechanism if and only if E is represented by a machine.*

Proof. We can regard any machine as a linear mechanism, the states of the machine becoming the nodes of the mechanism, and this does not affect the event represented. But for each linear mechanism (L, M, N) we can define a machine whose states are the subsets of the nodes α_i, with $\alpha_a = \{\alpha_j | a \in M_{ij} \text{ for some } i \text{ with } \alpha_i \in \alpha\}$, output function $o(\alpha) = 1$ if and only if there is $\alpha_j \in \alpha$ with $L_j = 1$, and initial state the set of all α_i with $L_i = 1$. This machine has 2^n states if (L, M, N) has n nodes, and represents the same event as (L, M, N).

The notion of linear mechanism is closely related to various notions of *non-deterministic automaton*. It is not quite the same, since we allow states in which no mode is active.

Theorem 7. *E is representable if and only if E can be expressed in the form $E = L(C + D)^* N$, where L, N are constant row and column vectors, and C, D are respectively constant and linear square matrices.*

Proof. The implication one way is immediate. Conversely, we observe that $(C + D)^* = (C^* D)^* C^*$, and so $L(C + D)^* N = L.(C^* D)^*.C^* N$, and that $(L, C^* D, C^* N)$ defines a linear mechanism, since $C^* D$ is linear and $C^* N$ constant.

We now prove the second half of Kleene's theorem by observing:

Theorem 8. *Any regular event is representable in the form of Theorem 7.*

Proof. Inductively by means of the formulae

$M2 \quad 0 = 0(0)^* 0$

$M3 \quad 1 = 1(1)^* 1$

$M4 \quad a = (1 \quad 0) \begin{pmatrix} 0 & a \\ 0 & 0 \end{pmatrix}^* \begin{pmatrix} 0 \\ 1 \end{pmatrix}$

$M5 \quad LM^* N + PQ^* R = (L \quad P) \begin{pmatrix} M & 0 \\ 0 & Q \end{pmatrix}^* \begin{pmatrix} N \\ R \end{pmatrix}$

$M6 \quad LM^* N.PQ^* R = (L \quad 0) \begin{pmatrix} M & NP \\ 0 & Q \end{pmatrix}^* \begin{pmatrix} 0 \\ R \end{pmatrix}$

$M7 \quad (LM^* N)^* = (0 \quad 1) \begin{pmatrix} M & N \\ L & 0 \end{pmatrix}^* \begin{pmatrix} 0 \\ 1 \end{pmatrix},$

which follow from $M1$ and show that 0, 1, and the inputs have such expressions, and that $E + F$, EF, E^* have them if E and F do.

For example, we take the event $E = (a^* b + bb^* a)^*$. $a^* b$ is represented by the two-node mechanism of Fig. 3.3, and $bb^* a$ by the three-node mechanism,

Fig. 3.3.

and so

$$a^*b + bb^*a = (1 \quad 0 \mid 1 \quad 0 \quad 0) \begin{bmatrix} a & b & 0 & 0 & 0 \\ 0 & 0 & 0 & 0 & 0 \\ \hline 0 & 0 & 0 & b & 0 \\ 0 & 0 & 0 & b & a \\ 0 & 0 & 0 & 0 & 0 \end{bmatrix}^* \begin{bmatrix} 0 \\ 1 \\ \hline 0 \\ 0 \\ 1 \end{bmatrix}$$

by $M5$, and so by $M7$,

$$E = (0 \quad 0 \quad 0 \quad 0 \quad 0 \mid 1) \begin{bmatrix} a & b & 0 & 0 & 0 & 0 \\ 0 & 0 & 0 & 0 & 0 & 1 \\ 0 & 0 & 0 & b & 0 & 0 \\ 0 & 0 & 0 & b & a & 0 \\ 0 & 0 & 0 & 0 & 0 & 1 \\ \hline 1 & 0 & 1 & 0 & 0 & 0 \end{bmatrix}^* \begin{bmatrix} 0 \\ 0 \\ 0 \\ 0 \\ 0 \\ \hline 1 \end{bmatrix}.$$

Thus E is represented by the mechanism of Fig. 3.4, from which we construct the machine of Fig. 3.5 (omitting inaccessible states). But in this machine $\{\alpha, \beta, \gamma, \zeta\}$ and $\{\alpha, \gamma, \epsilon, \zeta\}$ have the same output and are

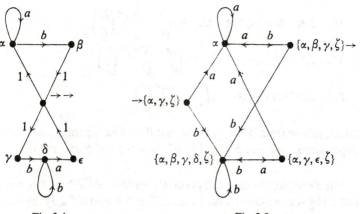

Fig. 3.4.   Fig. 3.5

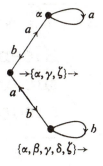

Fig. 3.6.

taken by a and b to the same state, and so are indistinguishable. Then these and $\{\alpha, \gamma, \zeta\}$ are indistinguishable, and the machine reduces to Fig. 3.6, which is isomorphic with that of Fig. 3.1.

We have cheated by calling Fig. 3.4 a mechanism. Really, it is a (*constant + linear*)-*mechanism*, defined as an ordered triple $(L, C + D, N)$ where these are as in Theorem 7. In this sort of mechanism, the activity of a node α_i instantaneously causes the activity of any α_j for which $C_{ij} = 1$. The (constant + linear)-mechanism $(L, C + D, N)$ is therefore equivalent to the linear mechanism (LC^*, DC^*, N), which is equivalent to a machine by Theorem 6.

Theorem 9. *There is an effective decision procedure enabling us to determine whether two regular expressions $R(a, b, c, \ldots)$ and $S(a, b, c, \ldots)$ represent the same event.*

Proof. Construct two binary machines M and N representing R and S, and then $R = S$ if and only if $M^{-\div}$ and $N^{-\div}$ are isomorphic.

We remark that our proof of Kleene's Theorem provides us with a fairly good method for passing from machines to regular expressions, but a deplorably inefficient one for the reverse process. The differential calculus of events will later improve this situation.

CHAPTER FOUR

Kleene algebras: the one-variable theorem

We introduce five different notions of *Kleene algebra*. The reason for this multiplicity is essentially that the formal laws satisfied by the regular operations are not easily codified:

An **S**-*algebra* (*standard* Kleene algebra) has already been defined.

An **N**-*algebra* (*normal* Kleene algebra) is any subset of an **S**-algebra which is closed under the regular operations +, ., * and contains 0 and 1, together with the operations +, ., *.

An **R**-*algebra* (*regular* Kleene algebra) is any set with two special elements 0 and 1 and operations +, ., * which satisfy all the formal laws which are universally valid in all **S**-algebras (or equivalently, in all **N**-algebras).

A **C**-*algebra* (*classical*, or *cyclic* Kleene algebra) is any set with special elements 0 and 1 and operations +, ., * which obey $C1$–14.

An **A**-*Algebra* (*acyclic* Kleene algebra) is a set with special elements 0 and 1 and operations +, ., * which obey $C1$–13.

For **X** = **S, N, R, C, A** (and later for certain other values of **X**) we say that an **X**-*algebra* has **X**-*operations* which satisfy all the **X**-*tautologies* between **X**-*expressions*. For **X** = **S** the **X**-operations are $\sum$, ., *, and the **X**-expressions are formulae built up by these operations from symbols for elements of an **X**-algebra. The **X**-tautologies are all equations $f(x,y,z,\ldots) = g(x,y,z,\ldots)$ between **X**-expressions which are universally valid in all **X**-algebras. For

$X = N, R, C, A$ the same holds except that the X-operations are $+, \cdot, *$.

It is obvious that every X-algebra is a Y-algebra for (X, Y) any of the pairs (S, N), (N, R), (R, C), (C, A), but we shall show later that this fails for each of the reverse pairs.

However, it is true that most of the simplest properties of X-algebras (for any X) are consequences of $C1$–14, and so we shall first investigate the consequences of these axioms. We shall not use $C13$ or $C14$ without explicit mention.

C15 $A^* = 1 + AA^*$ $(B = 1$ in $C12)$

C16.n $A^* = A^{<n} + A^n A^*$ (repeated use of $C15$)

C17 $(AB)^* A = A(BA)^*$ (by $C12$ $A(BA)^* = A + AB(AB)^* A = (AB)^* A$ by $C15$)

C18.n $A^{n*} A = AA^{n*}$ (for $n > 0$ put $B = A^{n-1}$ in $C17$ – trivial for $n = 0$)

C19 $0^* = 1$ $(A = 0$ in $C15)$

C13° $1^* = 1$ (deduced from $C13$ by putting $A = 0$)

So $C13$ implies $C13°$ in the presence of $C1$–12.

C21 $1 + 1 = 1$ $(A = 1$ in $C15$, then $C13°$; deduced from $C13°)$

C22 $A + A = A$ (multiply $C21$ by A; deduced from $C13°)$

C23 $(1 + A)^* = A^*$ $(B = A$ and $A = 1$ in $C11$, then deduced from $C13°)$

C13 $A^{**} = A^*$ $(A^* = (1 + A)^* = (A + 1)^* = A^*(A^*)^*$ by $C23$, $C2$, $C11$. Then $A^* = 1 + AA^* = 1 + 1 + AA^* = 1 + A^* = 1 + A^*(A^*)^* = A^{**}$, by $C15$, $C21$, $C15$, the previous line, and $C15$; deduced from $C13°)$

So $C13$ and $C13°$ are equivalent in the presence of $C1$–12.

C24 $A^* A^* = A^*$ $(B = 1$ in $C11$, $C13$, $C23$; deduced from $C13)$

It is worthy of note that all the idempotency laws $A + A = A$, $A^* A^* = A^*$, $A^{**} = A^*$ are consequences of the weak form $C13°$ of $C13$.

Definition. $A \leqslant B$ if and only if $A + B = B$.

C25 $0 \leqslant A$ $(C1, C2)$

C26 $A \leqslant A$ $(C22$; deduced from $C13)$

C27 $A \leqslant A + B$ $(C3, C22$; deduced from $C13)$

C28.n $A^{<n} \leqslant A^*$ $(C27, C16.n$; deduced from $C13)$

Theorem 1. (i) $A \leqslant B$ and $B \leqslant C$ imply $A \leqslant C$

(ii) $A \leqslant B$ and $B \leqslant A$ imply $A = B$

(iii) $A \leqslant B$ implies $A + C \leqslant B + C$

(iv) $A \leqslant B$ implies $AC \leqslant BC$

(v) $A \leqslant B$ implies $CA \leqslant CB$

(vi) $A \leqslant B$ implies $A^* \leqslant B^*$.

Proof. It should be noted that these follow from $C1$–12 alone, except that (vi) needs $C13$, or at least $C21$.

(i) uses $C3$, (ii) $C2$, (iii) is trivial, (iv) and (v) use $C8$ and $C9$, and for (vi) we have $A \leqslant B$ implies $A^* = 1A^* \leqslant (A^*B)^* A^* = (A + B)^* = B^*$.

$C29 . f \quad f(A, B, C, \ldots) \leqslant (A + B + C + \ldots)^*$ for every **X**-expression f.

Proof. Let $I = A + B + C + \ldots$, so that if V is any one of $A, B, \ldots$ we have $0 \leqslant V \leqslant I \leqslant I^*$, so that it suffices to prove that from $E \leqslant I^*$ and $F \leqslant I^*$ we may deduce $E + F \leqslant I^*$, $EF \leqslant I^*$, $E^* \leqslant I^*$. But these follow from $I^* + I^* = I^*I^* = I^{**} = I^*$, which are consequence of $C13$.

In what follows we shall use $C13$ and $C14$ freely. Our aim is to prove that every one-variable **R**-tautology is a **C**-tautology.

Theorem 2. *Suppose E is a finite sum of terms of the form $A^m A^{n*}$, with some $n > 0$, then $E^* \leqslant E^{<N}$ is a **C**-tautology for some N.*

Proof. Without loss of generality we may suppose that g, the greatest common divisor of all the numbers m and n, is 1, for otherwise we treat E as a function of A^g rather than of A.

Since $(1 + E)^* = E^*$ and $(1 + E)^{<N} = E^{<N}$ we can drop terms with $m = n = 0$, and then since $A^{m+n*} = A^{m+0*} + A^{m+n+n*}$ we may suppose that any term with $n > 0$ has $m > 0$ also. Now every term has $m > 0$, and so we have $E = AF$, say, and so $E \leqslant AA^*$ by $C29$, whence

$$E^N E^* \leqslant (AA^*)^N A^* = A^N A^*.$$

Now we shall show that for some N we have

$$A^N A^* \leqslant E^{<N}, \tag{1}$$

from which we can deduce

$$E^* = E^{<N} + E^N E^* \leqslant E^{<N} + A^N A^* \leqslant E^{<N} + E^{<N} = E^{<N} \leqslant E^*$$

and the conclusion of the theorem.

Now since $g = 1$ every sufficiently large integer is the sum of positive multiples of all the numbers m and n. But if $A^m A^{n*}$ is any term, and $rs > 0$, we have

$$E^r \geqslant (A^{m+n*})^r = A^{mr+n*} \geqslant A^{mr+ns}. \qquad (2)$$

Multiplying one inequality (2) for each term, and remembering that some term has $n > 0$, we see that for all sufficiently large integers P there is $Q < P$ with $E^Q \geqslant A^P$. Summing this for $P = N - m_0 + i$ ($0 \leqslant i < n_0$), where $A^{m_0} A^{n_0 *}$ is a term with $n_0 > 0$, we obtain

$$E^{<(N-1)} \geqslant A^{N-m_0 + <n_0}$$

whence

$$E^{<N} \geqslant E^{<(N-1)} E \geqslant A^{N-m_0 + <n_0} . A^{m_0 + n_0 *} = A^N A^{<n_0 + n_0 *} = A^N A^*,$$

using $C14.n_0$, and this is (1).

Theorem 3. *Any* **C**-*expression in the single variable A can be expressed as a finite sum of terms $A^m A^{n*}$, the empty sum being interpreted as zero.*

Proof. This is obviously true for 0 and for A, and true for $E + F$ if it is true for E and F. For EF we consider the product $A^m A^{n*} . A^M A^{N*}$. If n or N is zero, we use $C18$. If not, we can use $C14$ to express each of A^{n*} and A^{N*} in terms of A^{nN*}, and then use $C18$ and $C24$. So we consider an expression E^*, where E is already in the desired form. If $m = n = 0$ for each term, then E is 0 or 1, and so $E^* = 1$ by $C19$ and $C13°$. Otherwise, if $n = 0$ for each term we have $E = \sum A^m$ with some $m > 0$, and we can write $E = A^m + X$, and so $E^* = (A^m + X)^* = (A^{m*} X)^* A^{m*}$. Otherwise some term has $n > 0$, and the result follows from Theorem 2.

Theorem 4. *Any* **C**-*expression in the single variable A can be expressed in the form $p(A) + A^m q(A) A^{n*}$ in which $p(A)$ and $q(A)$ are polynomials (finite sums of powers of A) of degrees $<n$, $<m$, respectively. Any two such expressions can be put into such forms with the same value of n.*

Proof. We start with the expression of Theorem 3, and note that since $A^{n*} = (A^n)^{<k} \cdot A^{nk*}$ if $k > 0$, we can replace any n by nk, and so suppose that all the non-zero numbers n are equal. We can then use the tautology $A^{n*} = 1 + A^n A^{n*}$ to increase any m by the corresponding n, and so we have a form $p(A) + r(A)A^{n*}$ in which we can suppose that the index of any power of A in $r(A)$ exceeds that of any power of A in $p(A)$, so that we can write $r(A) = A^m q(A)$ for some m greater than the degree of $p(A)$. Since we can later replace n by any multiple of n, we can deal with two expressions simultaneously.

We can now suppress repeated powers of A in $p(A)$ or $q(A)$ in the canonical form of Theorem 4, and rearrange the terms of these polynomials in increasing order of indices. With such understandings, any **C**-expression has at most one canonical form for given numbers m, n.

Theorem 5. *Any* **R***-tautology in a single variable A is already a* **C***-tautology.*

Proof. Let the tautology be $f_1(A) = f_2(A)$, where f_1 and f_2 are in canonical form with the same m, n, and consider the regular events $f_1(a)$ and $f_2(a)$, where a is an input. If A^k appears in p_1 but not p_2, or in q_1 but not q_2, then a^k or a^{k+m} is in $f_1(a)$ but not $f_2(a)$, and so $f_1(a) \neq f_2(a)$. But if $f_1(A) = f_2(A)$ is an **R**-tautology, it must hold in every **S**-algebra, in particular, in the algebra of events over a single input a.

The reader might like to try his skill with the examples $(A^2 + A^3)^* = 1 + A^2 A^*$ and $A^4 + A^7 + (A^3 + A^5)^* = A^5 + (A^3 + A^4)^*$, both of which, being **R**-tautologies, must be deducible from $C1$–14.

We can express Theorem 5 in another way by saying that the free **C**-algebra in one generator is already a free **R**-algebra in that generator, these concepts being defined in the obvious way. We call an **X**-algebra generated by $x, y, z, \ldots$ *free* in those generators if and only if the equation $f(x, y, z, \ldots) = g(x, y, z, \ldots)$ holds for **X**-expressions f, g only when $f(X, Y, Z, \ldots) = g(X, Y, Z, \ldots)$ is an **X**-tautology. (In other words, when the only relations between the generators are those which hold in all **X**-algebras.) Then theorem 5 and its proof give a fairly complete structural characterization of the free **C**-algebra, hence the free **R**-algebra, in one generator.

We use $\mathbf{X}\langle a,b,...\rangle$ for the free **X**-algebra in generators $a, b, ...$ It is well known that this algebra, given that it exists, is unique to within isomorphism.

Theorem 6. $\mathbf{S}\langle a,b,...\rangle$ *is isomorphic with the* **S**-*algebra of* all *events in inputs* $a, b, ...$ $\mathbf{R}\langle a,b,...\rangle$ *and* $\mathbf{N}\langle a,b,...\rangle$ *are both isomorphic with the algebra of* regular *events in those inputs.*

Proof. Let $\bar{S}$ be any **S**-algebra generated by elements $\bar{a}, \bar{b}, ...$ For each word w in $a, b, . . .$, define $\bar{w}$ in $\bar{S}$ to be the product of the corresponding elements $\bar{a}, \bar{b}, . . .$, and for each event E in $a, b, . . .$ define $\bar{E} = \sum \bar{w}$ over $w \in E$. The map so defined is a homomorphism from the **S**-algebra of events in $a, b, . . .$ onto $\bar{S}$, and so no relation can hold for events in $a, b, . . .$ which does not hold for elements $\bar{a}, \bar{b}, . . .$ in all **S**-algebras. If we restrict ourselves to relations involving only the **R**-operations we get the second half of the theorem.

Corollary. $f(X, Y, Z,...) = g(X, Y, Z,...)$ *is an* **R**-*tautology if and only if the events* $f(a,b,c,...)$ *and* $g(a,b,c,...)$ *in distinct inputs* $a, b, c, ...$ *are equal. In view of Theorem 9 of the previous chapter this yields an effective test for* **R**-*tautologies.*

It was not *a priori* obvious that there existed a free **S**- or **N**-algebra in given generators, since the operation $\sum$ has unbounded scope (there is no proper concept of a free complete Boolean algebra in infinitely many generators). Theorem 6 justifies both concepts. In view of the Theorem, we refer to any member of $\mathbf{X}\langle a,b,...\rangle$ as an **X**-*event*, even for values of **X** not mentioned in the theorem and *infinite* sets $\{a,b,...\}$. **S**-*events* are arbitrary events, and **R**-*events* (equally **N**-events) are just regular events, but **C**-*events* and **A**-*events* will remain rather intangible concepts for some time.

Algebraic theorems about Kleene algebras are proved differently for the various values of **X**. For **S** we often have a simple abstract proof, and for **R** or **N** we can usually 'transfer' from the case **X** = **S**. But for **C** or **A** we must use complicated deductions from $C1$–14 without reference to any natural interpretations.

A good example is that the $n \times n$ matrices over an **X**-algebra form an **X**-algebra. For **X** = **S** this is trivial, and for **X** = **N** it is an immediate consequence of Theorem 5 of Chapter 3. For **X** = **R** we can deduce it as follows – we *define* the star of a matrix by repeated use of formula $M1$ (Chapter 3), and interpreting any **R**-tautology between $n \times n$ matrices as n^2 equations between their entries, we see that these equations must hold in all **S**-algebras, and so be **R**-tautologies, so that the original **R**-tautology holds for $n \times n$ matrices over any **R**-algebra (and the different ways of defining the star all give the same result). For **X** = **A** the theorem can be proved by explicit matrix computations, but for **X** = **C** it really begins to need the complicated techniques we apply to it in Chapter 13.

CHAPTER FIVE

The differential calculus of events

The theory developed in this chapter gives us an efficient technique for constructing binary machines to represent given events.

If a is any input, we define the *input derivate* $\partial E / \partial a = \partial_a[E] = E_a$ of the event E to be the set $\{w | aw \in E\}$. We define the *constant part* or *output* of E to be the event $o[E]$ which is 1 if $1 \in E$, and 0 otherwise. For E a regular function $f(a, b, c, \ldots)$ of inputs, $o[E] = f(0, 0, \ldots)$.

These notions satisfy:

$D0 \quad E = o[E] + aE_a + bE_b + \ldots (I = \{a, b, \ldots\})$

$D1 \quad o[0] = o[a] = 0 \ (a \in I) \qquad D5 \quad 0_a = b_a = 0, a_a = 1 \ (a, b \in I, a \neq b)$

$D2 \quad o[E + F] = o[E] + o[F] \qquad D6 \quad (E + F)_a = E_a + F_a$

$D3 \quad o[E.F] = o[E].o[F] \qquad D7 \quad (E.F)_a = E_a.F + o[E].F_a$

$D4 \quad o[E^*] = 1 \qquad D8 \quad (E^*)_a = E_a.E^*$

which provide inductive constructions for them, and show them to be regular for regular E. $D0$ is both Taylor's theorem and the mean value theorem in this theory.

If w is any word, say $w = ab \ldots k$, then we define the *word derivate* $\partial E / \partial w = \partial w[E] = E_w$ by iteration, $E_w = (\ldots(E_a)_b \ldots)_k$.

In any binary machine M, we define $E(\alpha)$ to be the set of all words w for which $o(\alpha_w) = 1$.

Theorem 1. *In any binary machine, $E(\alpha_a) = (E(\alpha))_a$ and $o(\alpha) = o(E(\alpha))$.*

Proof. We have $o(\alpha_{aw}) = 1$ if and only if $o((\alpha_a)_w) = 1$, and $1 \in E(\alpha)$ if and only if $o(\alpha_1) = 1$.

Now let $E = a^*b + bb^*a$. We shall compute all the word derivates of E^*. Using D1–8 we see, for instance, that $(E^*)_a = a^*bE^* = F$, say, and then that $F_a = F$ also, for $b_a = 0$, and $F_b = E^*$, using $(a^*)_b = 0$, $b_b = 1$. It can similarly be shown that $(E^*)_b = (1 + b^*a)E^* = G$, say, and that $G_a = E^*$, $G_b = G$. So every word derivate of E^* is one of E^*, F, G. We can summarize this information very neatly as in Fig. 5.1, where as arrow marked a is drawn from an event A to an event B whenever $A_a = B$, and an outgoing arrow $(A{\rightarrow})$ from an event A means that $o[A] = 1$. We have also attached an ingoing arrow $({\rightarrow}E^*)$ to our initial event E^*. Might there perhaps be some connection with machines?

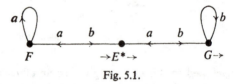

Fig. 5.1.

Theorem 2. *Let $M(E)$ denote the machine whose states are the word derivatives of E, with transition function $F_a = \partial F/\partial a$, output function $o(F) = o[F]$, and initial state E. Then $M(E)$ represents E.*

Proof. Repeated use of Theorem 1.

The implied algorithm is extremely efficient. The reader should follow the steps involved in the construction of the machine of Fig. 5.2

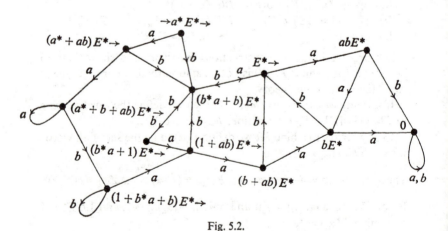

Fig. 5.2.

for the event $a*(a^2 b + bb* a + b^2)*$ – in this case, as in many others, the process gives the minimal machine directly to anyone skilled in input differentiation. The skill is worth acquiring, and the example will help the reader to acquire it. We write $E = a^2 b + bb* a + b^2$.

We can use these ideas to give another proof of the easy half of Kleene's main theorem only if we can find an independent proof that regular events have only finitely many word derivates. To find such a proof we generalize differentiation still further – for any events E, G we define the *event derivate* $\partial E/\partial G = \partial G[E] = E_G$ as the sum $\sum E_w$ over all $w \in G$, or equivalently as the set $\{v|Gv$ intersects $E\}$. The extended notion satisfies:

D9 $0_G = 0,\ 1_G = o[G]$

D10 $a_G = 0, 1, a,$ or $1 + a\ (a \in I)$ D14 $E_{F+G} = E_F + E_G$

D11 $(E + F)_G = E_G + F_G$

D12 $(E.F)_G = E_G.F + F_{G_E}$ D15 $E_{FG} = (E_F)_G$

D13 $(E*)_G = o[G] + E_{G_{E*}}.E*$

(D9 and D10 are trivial, and since the others (except D13) are in an obvious sense linear it suffices to prove them for *words e, f, g*. Then, for example, $(ef)_g = \{w|gw = ef\}$, and $gw = ef$ only if for some word u we have $e = gu, w = uf$ or $g = eu, f = uw$, which proves D12, for in the first case $u = e_g$, so $w = e_g.f$, and in the second case $u = g_e$, and $w = f_{g_e}$. By repeated use of D12 we can prove for $n > 0$ that $(E^n)_G$ is the finite sum $\sum E_{G_{E^r}}.E^s$ over all pairs (r,s) with $r + s = n - 1$, and D13 follows on summing this, and adding the term with $n = 0$.)

Theorem 3. *A regular event E has only finitely many distinct event derivates E_G (even for irregular G), and these are all regular events.*

Proof. Define the *length* $l(E)$ as the number of distinct derivates of E. Then D9 and D10 show that $l(0) = 1, l(a) = 4\ (a \in I)$, and D11–13 show that $l(E + F) \leqslant l(E).l(F),\ l(E.F) \leqslant l(E).l(F),\ l(E*) \leqslant 2l(E)$.

Another estimate of the complexity of E is its *width* $w(E)$, the number of word derivates of E, and a third is its *breadth* $b(E)$, the number of nodes in the smallest linear mechanism representing E.

Theorem 4. $b(E) \leqslant w(E) \leqslant l(E) \leqslant 2^{b(E)}$.

Proof. The machine of Theorem 2 can be regarded as a linear mechanism representing E – since it has $w(E)$ nodes we have $b(E) \leqslant w(E)$. Since every word derivate is an event derivate we have $w(E) \leqslant l(E)$. If (L, M, N) is a smallest linear mechanism representing E, then any derivate of E is represented by some linear mechanism (L_1, M, N), and since there are $2^{b(E)}$ values for L_1 we have $l(E) \leqslant 2^{b(E)}$.

We can also apply these measures to certain events related to E, notably the complement $-E$ of E and the *reverse* or *transpose* $\leftarrow E$ of E, obtained by reversing each word of E.

Theorem 5. *We have* $b(E) = b(\leftarrow E)$, $w(E) = w(-E)$, $l(E) = l(\leftarrow E)$.

Proof. If (L, M, N) represents E, then (N', M', L') represents $\leftarrow E$, the primes indicating transposed matrices. For any word w we have $(-E)_w = -(E_w)$. These remarks prove $b(E) = b(\leftarrow E)$, $w(E) = w(-E)$. To prove $l(E) = l(\leftarrow E)$ we observe that for any event G we have $w \notin E_G$ if and only if wG is disjoint from E, so that $-E_G$ is the largest event F for which FG is disjoint from E. Having determined this F, we can find the largest H with FH disjoint from E, and then plainly will $F = -E_H$. So if we find all pairs (L, R) in which L and R are maximal subject to LR being disjoint from E, then the derivates of E will be the events $-L$, and those of $\leftarrow E$ the events $-(\leftarrow R)$, and the two systems are in 1–1 correspondence. Amplifications appear in the next chapter.

We note that an event over a finite input population is regular if and only if it has finitely many derivates, or equally, if and only if it has finitely many word derivates. Over an infinite input population this is no longer true – the events with finitely many derivates are just those which can be expressed as regular functions of (arbitrary) sums of inputs. In particular, the universal event I^* is not regular for infinite I, but has only one derivate (itself).

Theorem 6. *The class of regular events over a finite input population is closed under the Boolean operations. In other words, if E and F are regular, so are $E \cap F$, $E \setminus F \, (= E \cap (-F))$, $-E$, etc.*

Proof. For any w we have $(E \cap F)_w = E_w \cap F_w$, $(E \backslash F)_w = E_w \backslash F_w$, and $(-E)_w = -E_w$, and so $E \cap F$, $E \backslash F$ and $-E$ have only finitely many word derivates if E and F have.

We consider for example the expression $E^* \cap F^*$, wherein $E = a^*b + bb^*a$ and $F = a^2 + b^2$. We obtain the machine of Fig. 5.3

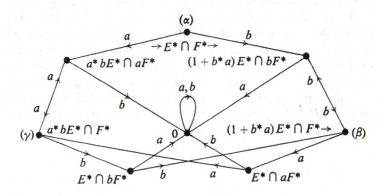

Fig. 5.3.

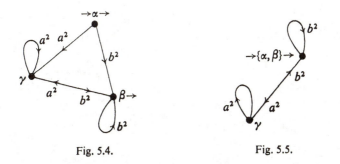

Fig. 5.4.   Fig. 5.5.

by differentiating this expression as it stands, noting only that any derivate $A \cap 0$ or $0 \cap A$ can be replaced by 0. We then simplify by noting that the result must be a function of a^2 and b^2 only, giving Fig. 5.4, reducing to Fig. 5.5, from which we can read off the regular expression $(a^2 a^* b^2 + b^2)^*$, or more simply, $(a^2 * b^2)^*$.

Of course the objects depicted in Figs. 5.4 and 5.5 are not machines in the original sense, but yet another generalization. Let us define once and for all a *general mechanism* as a triple (L, M, N) wherein L is a row vector of length n, M an $n \times n$ matrix, and N a column vector of length n, L, M, N having arbitrary events for entries. In the diagrammatic representation we have n nodes, the ith of which has an ingoing arrow marked L_i, an outgoing arrow marked N_i, and is joined to the jth by an arrow marked M_{ij}. We omit arrows marked zero, and the marks of arrows marked 1. The typical branch, marked A, say, is a kind of *transducer*: if its tail is active at some time and a word $w \in A$ is applied, then the head will become active. The transducer will presumably have components inside which contain all the relevant information about times at which the tail has been active. The general mechanism (L, M, N) *represents* the event $E = LM*N$. We could also (and this seems more natural) add a branch marked K going directly from the input to the output, and then the represented event would be $K + LM*N$. We leave the problem of defining general mechanisms more formally to the interested reader.

CHAPTER SIX

Factors and
the factor matrix

The type of problem we shall solve in this chapter is the following. How can we express a given regular event E in the form $f(F_1, F_2, \ldots)$, wherein f is a regular function and the F_i regular events? We might be given the function f and required to determine the F_i, as for instance in the case $F_1^2 = E$. Or we could be given the events F_i and want to express E as a function of them. This problem might arise in the construction of large machines from smaller ones – given transducers for the F_i, how can we build a transducer for E?

Definitions (in which the letters represent events):

$F.G \ldots H \ldots J.K$ is a *subfactorization* of E if and only if

$$FG \ldots H \ldots JK \leqslant E. \qquad (*)$$

$\bar{F}.\bar{G} \ldots \bar{H} \ldots \bar{J}.\bar{K}$ *dominates* it if $F \leqslant \bar{F}, G \leqslant \bar{G}, \ldots, K \leqslant \bar{K}$.
A term H is *maximal* if it cannot be increased without violating (*).
A *factorization* of E is a subfactorization with every term maximal.
A *factor* of E is any event which is a term in some factorization of E.
A *left* (*right*) factor is one which can be the leftmost (rightmost) term in a factorization. Note that the equation $LR = E$ is neither implied by, nor implies, that $L.R$ is a factorization of E.

Theorem 1. *Any subfactorization of E is dominated by some factorization in which all terms originally maximal remain unchanged.*

47

Proof. In the subfactorization (*), replace G, say, by $\bar{G}$, the sum of all events G_1 for which $FG_1 \ldots H \ldots JK \leqslant E$. We obtain a subfactorization in which $\bar{G}$ is maximal as well as any terms which were already so. Repeating the process with the remaining terms in any order we eventually obtain a factorization.

The factorization so obtained need not be unique – indeed, all factorizations of E dominate the subfactorization $0.0.0\ldots 0$. But the argument shows it unique if only one term was originally not maximal.

Theorem 2. *Any left factor is the left factor in some 2-term factorization. Any factor is the central term in some 3-term factorization. Any right factor is the right factor in some 2-term factorization.*

Proof. If (*) is a factorization then F, H, K are maximal in the sub-factorizations $F.(G\ldots K)$, $(FG\ldots).H.(\ldots JK)$, $(FG\ldots).K$ respectively. Apply Theorem 1.

Theorem 3. *The condition that $L.R$ be a factorization of E defines a 1–1 correspondence between the left and right factors.*

Proof. The previous theorems show that each L corresponds to at least one R, and vice versa – but for given L the sum of all possible R is the only possible maximal R.

Notation. We index the left and right factors as L_i, R_i so that $L_i.R_i$ is a factorization, and define E_{ij} by the condition that $L_i.E_{ij}.R_j$ be a subfactorization of E in which E_{ij} is maximal (this is possible, since $L_i.0.R_j$ is a subfactorization).

Theorem 4. *Each E_{ij} is a factor, and each factor is one of the E_{ij}. There exist unique indices l, r such that $E = L_r = R_l = E_{lr}$ and $L_i = E_{li}$, $R_i = E_{ir}$ for each i. Hence the factors naturally form a square matrix among the entries of which is E.*

Proof. E_{ij} is maximal in $L_i.E_{ij}.R_j$, and so is a factor by Theorem 1. Each factor H is the central term in a 3-term factorization $L_i.H.R_j$ by Theorem 2, and so we must have $H = E_{ij}$ for some i, j, not necessarily unique. The term E is maximal in the subfactorization $1.E$, dominated by $L_l.E$, say, whence $E = R_l$ and $L_l \geqslant 1$. Now for any i,

$L_i.L_i.R_i$ is a subfactorization with L_i, R_i maximal, and so $L_i = E_{ii}$. Defining r similarly by the condition that $E.R_r$ be a factorization, we find that $L_i E \leqslant E$ and $ER_r \leqslant E$, so $L_i ER_r$ is a subfactorization with E maximal, whence $E = E_{lr}$.

Note. This organization of the factors as E_{ij} has many remarkable properties. The theorem does prevent E from occurring twice in its factor matrix, and in general certain factors appear repeatedly. It follows easily from later theorems that we have $l = r$ if and only if E is a perfect star.

Theorem 5. *An event E over a finite input population has finitely many factors if and only if it is regular. The number of left or right factors of E is $l(-E)$. The factors are regular for regular E.*

Proof. The right factors R of E are just the maximal events with $LR \leqslant E$, L ranging over all events. But $LR \leqslant E$ if and only if $wR \leqslant E$

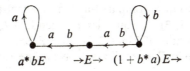

$$a^* bE \qquad \rightarrow E \rightarrow \qquad (1 + b^* a)E \rightarrow$$

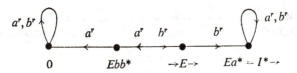

$$0 \qquad Ebb^* \qquad \rightarrow E \rightarrow \qquad Ea^* = I^* \rightarrow$$

Fig. 6.1. A machine and its antimachine.

for each $w \in L$; that is to say, if and only if $R \leqslant \cap E_w$ (over $w \in L$). The set of right factors is therefore the set of all intersections of word derivates, and the first part of the theorem follows. Complementing, we obtain $-R = \sum (-E)_w$ (over $w \in L$), so that E's typical right factor is the complement of the typical derivate of $-E$, and the rest follows.

For an example, we compute the factor matrix of the event $E = (a^* b + bb^* a)^*$, and simultaneously introduce some new concepts.

The *right derivate*, or *antiderivate* $\partial G^r[E] = \partial E/\partial G^r$ is defined as the set $\{w|wg \text{ intersects } E\}$. Differentiating on the right we obtain an *antimachine* for E. Machine and antimachine for this particular E are shown in Fig. 6.1. The intersections of the states of the machine are the right factors, and the intersections of states of the antimachine are the left factors. In our case, we have

$$a^* bE \leqslant E \leqslant (1 + b^* a) E,$$

so that the right factors are $a^* bE \leqslant E \leqslant (1 + b^* a)E \leqslant I^* = (a + b)^*$ (I^* is the *empty* intersection). In a similar way the left factors are $0 \leqslant Ebb^* \leqslant E \leqslant I^*$, and we can write down most of the factor matrix at sight:

	0	1	2 = r	3
0	I^*	I^*	I^*	I^*
1	0	?	$(1 + b^* a) E$	I^*
$l = 2$	0	Ebb^*	E	I^*
3	0	?	$a^* bE$	I^*.

E_{11} and E_{31} are now computed as the maximal events such that $L_i E_{i1} R_1 \leqslant E$ ($i = 1, 3$), or equivalently, such that $L_i E_{i1} \leqslant L_1$, so we construct the machine of Fig. 6.2 for $L_1 = F$, say, and compute E_{i1} as the intersection of F_w over all $w \in L_i$. Now $L_1 = (a^* b + bb^* a)^* bb^*$,

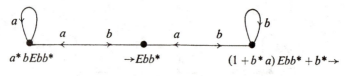

Fig. 6.2.

and it is easy to see that the general word of L_1 takes us from the initial state of this machine to the state $(1 + b^* a) Ebb^* + b^*$, so that E_{11} is this event, and similarly $E_{31} = a^* bEbb^*$, completing the matrix.

Theorem 6. *We have*

    (i) $1 \leqslant E_{ii}$;

    (ii) $E_{ij} E_{jk} \leqslant E_{ik}$;

    (iii) $AB \leqslant E_{ik}$ *if and only if* $A \leqslant E_{ij}$, $B \leqslant E_{jk}$ *for some j*;

    (iv) $F_a F_b \ldots F_k \leqslant E$ *if and only if there are indices l, m, n, . . ., q, r for which* $F_a \leqslant E_{lm}$, $F_b \leqslant E_{mn}$, . . ., $F_k \leqslant E_{qr}$ (*l, r as usual*).

Proof. (i) $L_i . 1 . R_i \leqslant E$, so $1 \leqslant E_{ii}$;

    (ii) $L_i E_{ij} R_j \leqslant E$, so $L_i E_{ij} \leqslant L_j$, whence $L_i E_{ij} E_{jk} R_k \leqslant L_j E_{jk} R_k$
       $\leqslant E$, whence $E_{ij} E_{jk} \leqslant E_{ik}$;

    (iii) If $AB \leqslant E_{ik}$, then $L_i A . B R_k \leqslant E$, and so $L_i A \leqslant L_j$, $B R_k \leqslant R_j$
      for some j, so that $L_i A R_j$ and $L_j B R_k \leqslant E$, whence $A \leqslant E_{ij}$,
      $B \leqslant E_{jk}$;

    (iv) inductively from (iii).

Corollary 1. *Factors of factors are themselves factors.*

Corollary 2. *The factor matrix $\overline{|E|}$ is its own star.*

It should not be thought, however, that any expression such as $L_i E_{ij} R_j$ is necessarily a factorization of E.

We shall soon see that the equation $\overline{|E|} = \overline{|E|}*$ gives us in a strong sense the most general expression for E in terms of events $F_1, \ldots, F_k$. Suppose for a moment that the F_t are given. Then among the possible functions $f(x_1, x_2, \ldots)$ for which $f(F_1, F_2, \ldots) \leqslant E$ there is a maximal one, namely the set of all words $x_a x_b \ldots x_k$ for which $F_a F_b \ldots F_k \leqslant E$. This function has the twin advantages that it is unique and defined for all systems E, F_1, F_2, $\ldots$, even when E is not expressible in terms of the F_t. We call it the *best approximating function* for E in terms of the F_t, and the event $f(F_1, F_2, \ldots)$ is the *best approximation* to E by the F_t.

In cases when E is expressible as a function of the F_t, we shall have $f(F_1, F_2, \ldots) = E$, but in other cases we shall of course have strict inequality. Note also that we can have $f(F_1, F_2, \ldots) = g(F_1, F_2 \ldots)$ for some function g other than f.

We can attempt to approximate E in more restricted ways. The *best linear approximating function* for E by the F_t is the sum of all the variables x_t for which $F_t \leqslant E$, and the *best linear approximation* is the sum of the corresponding F_t. The *best constant approximation* is of course the constant part $o[E]$. We approximate a matrix in any of these ways by approximating its entries individually.

Theorem 7. *Let F, C, L denote respectively the best, best constant, best linear approximating functions for $\overline{|E|}$ by the F_t. Then*

$$F = C + LL* = C + L* = (C + L)*.$$

Hence, the best approximation to $\boxed{E}$ by the F_t is the star of the sum of the best linear and best constant approximations.

Proof. The non-empty word $x_a x_b \dots x_k$ belongs to F_i if and only if $F_a, F_b, \dots, F_k$ satisfy the conditions of Theorem 6 part (iv), which are precisely the conditions under which it belongs to LL^*. The empty word belongs to F_{ij} if and only if $C_{ij} = 1$. These remarks prove the first equality, and the others are proved similarly, we have $CL = L = LC$.

Theorem 7 embodies an algorithm for computing best approximating functions. We first compute the factor matrix $\boxed{E}$, then replace every entry E_{ij} by the sum of all the x_t for which $F_t \leqslant E_{ij}$, together with 1 if $1 \in E_{ij}$, and then take the (l, r) entry in the star of the resulting matrix.

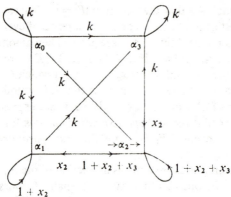

Fig. 6.3.

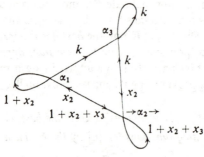

Fig. 6.4.

Let us consider the standard example $E = (a*b + bb*a)*$, and take $F_1 = a^2$, $F_2 = b$, $F_3 = ba$, the input set being $I = \{a,b\}$. Then the sum of the best constant and best linear approximating functions for $\overline{E}$ is

	0	1	2	3
0	k	k	k	k
1	0	$1 + x_2$	$1 + x_2 + x_3$	k
2	0	x_2	$1 + x_2 + x_3$	k
3	0	0	x_2	k

where $k = 1 + x_1 + x_2 + x_3$, so that the $(2,2)$-element of the star of this (say $f(x_1, x_2, x_3)$) will be the best approximating function for E by the F_t. It follows that $f(x_1, x_2, x_3)$ is the event represented by the general mechanism of Fig. 6.3, which we can simplify successively to Figs. 6.4, 6.5 and 6.6, getting finally $(x_2 + x_3 + x_1(x_1 + x_3)*x_2)*$ as the best approximating function, and $(b + ba + a^2(a^2 + ba)*b)*$ as the best approximation.

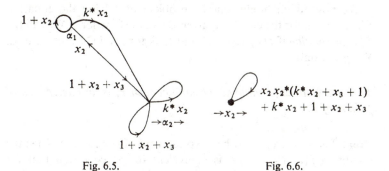

Fig. 6.5.   Fig. 6.6.

As corollaries to Theorem 7 we have:

Corollary 1. *There is an effective procedure which determines whether a given regular E is expressible in terms of given regular F_t.*

Corollary 2. *If E is regular any best approximating function for E is regular, even when the F_t are not.*

Corollary 3. *For a given number of events F_t, and a regular E, we obtain only finitely many approximating functions as the F_t vary.*

The factorial function. Let $E_1, E_2, \ldots, E_k$ denote the distinct factors of E, let $x_1, x_2, \ldots, x_k$ be distinct variables, and let M be the matrix obtained from $\overline{|E|}$ by replacing each E_t by the corresponding x_t. The *factorial function* $E!(x_1, \ldots, x_k)$ is defined as the sum of $o[E]$ and the (l, r)-element of M^*. This single function is a complete prescription for the structural makeup of E. We shall write $E = E!(E_1, \ldots, E_k)$ to mean at the same time that $E_1, \ldots, E_k$ are the distinct factors of E and that $E!(x_1, \ldots, x_k)$ the factorial function of E. In this notation, Theorem 7 simplifies:

Theorem 8. *If* $E = E!(E_1, \ldots, E_k)$, *then the best approximation to E by events $F_1, F_2, \ldots$ is $E!(Y_1, \ldots, Y_k)$, where Y_i is the best linear approximation to E_i by the F_t.*

We remark that it is seldom necessary to compute $\overline{|E|}$ in order to find the best approximations. It suffices to find the factors L_i, R_i, and then determine the matrices C and L as the maximal constant and linear matrices such that $L_i(C_{ij} + L_{ij}) R_j \leqslant E$ for each i, j.

We now touch briefly on the problem of solving the equation $f(F_1, \ldots) = E$ for the events F_t, given E and the function f. We call the F_t a *subsolution* if $f(F_1, \ldots) \leqslant E$, and look for maximal subsolutions. We prove only

Theorem 9. *In any maximal subsolution each F_t is an intersection of factors of E.*

Proof. Let $w = x_a x_b \ldots x_k$ be a typical word of $f(x_1, x_2, \ldots)$. Then the condition $f(F_1, F_2, \ldots) \leqslant E$ is equivalent to the condition that for every w there are indices $l, m, n, \ldots, q, r$ for which

$$F_a \leqslant E_{lm}, \; F_b \leqslant E_{mn}, \; \ldots, \; F_k \leqslant E_{qr},$$

l and r being as usual. If the F_t are a maximal subsolution and we increase any one of them there will exist a word w for which this fails. So the condition is equivalent to some one of a system of inequalities of the form

$$F_t \leqslant \text{some intersection of the factors of } E.$$

If the F_t are maximal we shall get equality in at least one of these conditions for each t.

Corollary. *We can, with patience, find all maximal subsolutions. For we have only to try all possible sets of values for the F_t as intersections of the E_{ij} and reject any non-maximal ones!*

In particular cases we can do better. Let us define a *square root* of E as any maximal A such that $A^2 \leqslant E$. Then, since $A^2 \leqslant E$ if and only if $A \leqslant L_i$ and $A \leqslant R_i$ for some i, all the square roots are found under the form $L_i \cap R_i$ and there are, at most, $l(-E)$ of them. The reader should show that the only square root of the event $E = (a(a+b)+b)^* a$ is $a^* bE$. In a similar way, each *cube root* of E has the form $L_i \cap E_{ij} \cap R_j$ for some i, j, and there are, at most, $(l(-E))^2$ cube roots.

Many other problems concerning equations in regular events can be tackled by similar methods supplemented by ad hoc techniques. The reader might like to prove that for any E there is a unique *minimal* A such that $A^* \geqslant E^*$, and that A (the *starth root* of E) is regular whenever E is. A problem we have not been able to solve concerns the *normalizer* $\mathcal{N}(E)$ of E, defined as the largest event F such that $EF = FE$. When E is regular, is $\mathcal{N}(E)$ regular?

Many theorems about intersections of regular functions are proved most easily by factor theory. We give as a sample:

Theorem 10. *If E is regular, then $E \cap AB$ can be expressed as a finite sum of terms $A_i B_i$ in which $A_i \leqslant A$, $B_i \leqslant B$. In particular, if $E \leqslant AB$, then E is a finite sum of such terms. If A and B are regular, then so are A_i and B_i.*

Proof. These statements follow at once from the formula

$$E \cap AB = \sum (L_i \cap A)\ (R_i \cap B).$$

CHAPTER SEVEN

The theory of operators: biregulators

Let S be any **S**-algebra. Then a function $\Omega: S \to S$ will be called an *S-operator*, and we shall use square brackets to contain the arguments of such operators. The *zero* and *identity* S-operators are defined by $0[X] = 0$, $1[X] = X$ for all $X \in S$. We then introduce algebraic operations on S-operators by defining $(\sum \Omega_t)[X] = \sum (\Omega_t[X])$, both sums being over the same index set, $\Omega\Psi[X] = \Omega[\Psi[X]]$, and $\Omega^* = 1 + \Omega + \Omega^2 + \ldots$.

Unfortunately, the set of all S-operators does not form an **S**-algebra under these operations, since the distributive law fails. Thus it is natural to restrict ourselves to the set $\langle S, S \rangle$ of **S**-*linear* S-operators, these being defined by the requirement that $\Omega[\sum X_t] = \sum \Omega[X_t]$ for all systems (X_t) of elements of S. In particular, $\Omega[0] = 0$ for an **S**-linear S-operator. If S is a *free* **S**-algebra, and E and F events in S, we define the operator $[E, F]$ by setting $[E, F][X] = E$ if F intersects X, 0 otherwise.

Theorem 1. *The set $\langle S, S \rangle$ of all **S**-linear S-operators is an **S**-algebra for any **S**-algebra S. If S is free, then every element of $\langle S, S \rangle$ is uniquely expressible as a sum of operators $[u, v]$ (u, v words), and these operators satisfy the relations $[u, v][v, w] = [u, w]$, $[u, v][v', w] = 0(v \neq v')$, which completely determine the structure of $\langle S, S \rangle$.*

Proof. The verification of the axioms for an **S**-algebra is immediate. Let the S be a free **S**-algebra, Ω any element of $\langle S, S \rangle$, and Ω' the sum

56

of all the operators $[u,v]$ belonging to a certain set Q. Then by linearity we have $\Omega = \Omega'$ if and only if $\Omega[v] = \Omega'[v]$ for all words v, and this holds only if Q consists exactly of the operators $[u,v]$ for which $u \in \Omega[v]$. Finally, the multiplicative relations for the operators $[u,v]$ are readily checked.

Convention. In view of the above result we can safely identify each $\Omega \in \langle S, S \rangle$ with the corresponding set Q of operators $[u,v]$, and so we shall write $[u,v] \in \Omega$ to mean $u \in \Omega[v]$. In particular,

$$[E, F] = \{[u,v] \mid u \in E, v \in F\}.$$

From now on we shall use the unqualified word 'operator' to mean '**S**-linear S-operator over a *free* **S**-algebra with a *finite* input alphabet $I = \{a, b, c, \ldots\}$'. Many of our theorems hold in more general circumstances, but it would be tedious to give exact conditions separately in each case.

We have already shown that the class of regular events is closed under a number of constructions – for instance under Boolean operations and under formation of derivates. One of the main aims of our theory of operators is the discovery of further closure properties of the regular events. Accordingly, we call the operator Ω a *regulator* if $\Omega[X]$ is regular for each regular event X, and a *total regulator* if $\Omega[X]$ is regular for *all* X. In this chapter we shall study more particularly the *biregulators* which we are about to define.

The biregular operations, and biregulators

We have shown that the set $\langle S, S \rangle$ is abstractly the set of all subsets of the Cartesian product set $S \times S$. This suggests that we consider the alternative multiplication defined by setting

$$\Omega : \Omega' = \{[uu', vv'] \mid [u,v] \in \Omega, [u',v'] \in \Omega'\}.$$

$\Omega : \Omega'$ is called the *biproduct* of Ω and Ω'. The operator $[1,1]$, which we call the *biunit*, is an identity for this multiplication, and so we can define the *bistar* of Ω by the equation

$$\Omega^{**} = [1, 1] + \Omega + \Omega : \Omega + \Omega : \Omega : \Omega + \ldots.$$

Plainly $\langle S, S \rangle$ is an **S**-algebra under these *biregular operations* $+$, $:$, $**$

as well as under the *regular operations* +, . , *. Of course, we must distinguish between the new meaning for Ω^{**} and the old meaning $(\Omega^*)^*$, but since $(\Omega^*)^* = \Omega^*$ the latter need never arise. If $f(a,b,c,\ldots)$ is a regular expression in its arguments, and $A, B, C, \ldots$ are operators, then we use $f((A,B,C,\ldots))$ for the operator obtained by replacing each

A table of biregulators

$[E,1][X] = E$ if $X \geqslant 1$	$[E,1] = f(([a,1],[b,1],\ldots))$
$[1,E][X] = 1$ if $X \cap E \neq 0$	$[1,E] = f(([1,a],[1,b],\ldots))$
$[E,F][X] = E$ if $X \cap F \neq 0$	$[E,F] = [E,1]:[1,F] = [1,F]:[E,1]$
$1[X] = X$	$1 = ([a,a]+[b,b]+\ldots)^{**}$
$E^{\cap}[X] = E \cap X$	$E^{\cap} = f(([a,a],[b,b],\ldots))$
$E^+[X] = E + X$ if $X \neq 0$	$E^+ = 1 + [E,I^*]$
$E^l[X] = EX$	$E^l = [E,1]:1 = [E,1)$ say
$E^r[X] = XE$	$E^r = 1:[E,1] = (E,1]$ say
$\partial E^l[X] = \dfrac{\partial X}{\partial E} = \{t \mid Et \cap X \neq 0\}$	$\partial E^l = [1,E]:1 = [1,E)$ say
$\partial E^r[X] = \dfrac{\partial X}{\partial E^r} = \{t \mid tE \cap X \neq 0\}$	$\partial E^r = 1:[1,E] = (1,E]$ say
$E\dfrac{\partial}{\partial F}[X] = E\dfrac{\partial X}{\partial F}$	$E\dfrac{\partial}{\partial F} = [E,F]:1 = [E,F)$ say
$E^r\dfrac{\partial}{\partial F^r}[X] = \dfrac{\partial X}{\partial F^r}.E$	$E^r\dfrac{\partial}{\partial F^r} = 1:[E,F] = (E,F]$ say
$\partial E^l[X] = \dfrac{\partial E}{\partial X}$	$\partial E^l = \sum [R_l,L_l]$
$\partial E^r[X] = \dfrac{\partial E}{\partial X^r}$	$\partial E^r = \sum [L_l,R_l]$
$(E,F)[X] = \{uev \mid uFv\ X \neq 0, e \in E\}$	$(E,F) = 1:[E,F]:1$
$E^{lr}[X] = \{uxv \mid x \in X, uv \in E\}$	$E^{lr} = \sum L_l^l:R_l^r$
$E^{l\leftarrow r}[X] = \dfrac{\partial E}{\partial(\leftarrow X)^r}$	$E^{l\leftarrow r} = \sum [L_l, \leftarrow R_l]$
$E\$[X] = E\$X$	$E\$ = 1:f((a^l,b^l,\ldots))$
$E\$\$[X] = E\$\$X$	$E\$\$ = f(([a,1]:I^{\cap},[b,1]:I^{\cap},\ldots))$
$x_+[X] = x^*g(ax^*, bx^*, \ldots)$	$x_+ = (([x^*,1])+1)^{**}$
$a_-[X] = g(1, b, \ldots)$	$a_- = ([a,1]+[b,1]+\ldots)^{**}$
$\Omega[X] = g(E,F,\ldots)$ when $X = g(a,b,\ldots)$	$\Omega = ([E,a]+[F,b]+\ldots)^{**}$

occurrence of any one of $a, b, c, \ldots, 1, +, . , *$ in $f(a,b,c,\ldots)$ by the corresponding one of $A, B, C, \ldots, [1,1], +, :, {**}$. Thus if $f(a,b,c) = ab^* + (ca)^*$ we have $f((A,B,C)) = A:B^{**} + (C:A)^{**}$. We call $f((A,B,C,\ldots))$ a *biregular function* of $A, B, C, \ldots$.

An operator is called a *biregulator*, or said to be *biregular*, if it can be expressed as a biregular function of the operators $[a,1], [1,a], [b,1]$,

$[1,b]$, ..., where a, b, ... are the input letters. There is an extensive theory of biregulators which we shall soon develop, showing in particular that every biregulator is a regulator, and that the product $\Omega\Psi$ of two biregulators is another (but the star of a biregulator need not even be a regulator). First we hint at the value of such a theory by listing some particular biregulators. We suppose that E, F, ... are regular events, and describe the action of various operators in one column, while in the other column giving a formula to prove the biregularity. We understand that $E = f(a, b, ...)$ where f is regular, and that $L_i . R_i$ is the typical 2-term factorization of E. When we write '$\Omega[X] = A$ if $P(X)$' we intend that $\Omega[X] = 0$ if $P(X)$ does not hold. Remember that the input alphabet $I = \{a, b, ...\}$ is *finite*.

The last few entries need some explanation. The *shuffle* $E \$ X$ of E and X is defined to be the set of all words of the form $e_1 x_1 e_2 x_2 \ldots e_n x_n$ (as n varies) for which e_1, x_1, e_2, x_2, ... are *words* such that $e_1 e_2 \ldots e_n \in E$, $x_1 x_2 \ldots x_n \in X$. The *alternate shuffle*, or *bishuffle*, $E \$\$ S$, is obtained if we further restrict e_1, ..., e_n, x_1, ..., x_n to be *letters*. We shall discuss the *expansion* and *contraction* operators a_+ and a_- in a moment. The operator $\Omega = ([E, a] + [F, b] + ...)^{**}$ is the typical regular *unit-homomorphism* – that is to say, it satisfies $\Omega[XY] = \Omega[X]\Omega[Y]$ and $\Omega[1] = 1$, and is a regulator. We intend all the operators in the table to be defined for general E, F, ..., although in general they will only be biregular for regular E, F, ...

The embedding of S in $\langle S, S \rangle$. Duality

It is easy to see that the map taking E to the corresponding left multiplier operator E^l is an isomorphism from S onto its image S^l in $\langle S, S \rangle$, which we call the algebra of left multipliers. It is convenient to identify S with S^l, or in other words, to abbreviate E^l to E for each event E. At the same time, we abbreviate ∂E^l to ∂E, and $_\partial E^l$ to $_\partial E$, but when there is risk of confusion we shall revert to the longer names. The map taking E to E^r is an *anti-isomorphism* onto its image, since we have $(EF)^r = F^r E^r$.

It follows from Theorem 1 that the algebra $\langle S, S \rangle$ has a natural anti-isomorphism – we write $\partial\Omega$ for the operator $\{[v, u] | [u, v] \in \Omega\}$, so that $v \in \partial\Omega[u]$ if and only if $u \in \Omega[v]$. Then we have $\partial\partial\Omega = \Omega$, $\partial(\Omega . \Psi) = \partial\Psi . \partial\Omega$, and we call Ω and $\partial\Omega$ *dual operators*. We see from the table that $E^\cap$ is self dual, while E, $\partial/\partial E$ and E^r, $\partial/\partial E^r$ are dual pairs, as are $^{\partial}E$, $^{\partial}E^r$ and $[E, F]$, $[F, E]$. Thus the operator $^{\partial}\Omega$ can be thought of as 'differentiation by Ω', and we might occasionally write $\partial/\partial\Omega$. Since $^{\partial}f((\Omega, \Omega', \ldots)) = f((^{\partial}\Omega, \Omega', \ldots))$, the dual of a biregulator is again a biregulator.

The reader should not think that every regulator is biregular. Let us define $\leftarrow$ by $\leftarrow[X] = \leftarrow X$, the set of all reverses of words of X, and $\bigcirc[X]$ as the set of all words obtained by cyclically rotating words of X. Then $\leftarrow$ and $\bigcirc$ are regulators, but we shall later show that neither is a biregulator. To see that $\bigcirc$ is a regulator, we need only observe that $\bigcirc[E] = \sum R_i . L_i$ in the usual notation. Note that $\bigcirc$ is the star of a biregulator – if $Q = [1, 1] + \sum {}^{\partial}a : a^r$ (the sum over all letters), then $Q[X]$ is the set of all words obtained by rotating words of X by one letter, and so $\bigcirc = Q^*$. Thus the star of a biregulator can be a regulator without being biregular. In fact, it need not even be a regulator – if $\Omega = [a^2, a]^{**}$ then Ω is biregular, and replaces a^n by a^{2n}, but $\Omega^*[a] = a + a^2 + a^4 + a^8 + \ldots$, an irregular event.

Theorem 2. *The class of regulators is closed under the biregular operations.*

Proof. Suppose that Ω and Ψ are regulators. Then, since $[1, 1]$ and $\Omega + \Psi$ are obviously regulators, we need only prove that $\Omega : \Psi[E]$ and $\Omega^{**}[E]$ are regular for each regular E. But it is immediate from the definitions that $\Omega : \Psi[E]$ is the sum of all expressions $\Omega[u] . \Psi[v]$ for which $uv \in E$, and so from factor theory is equal to the finite sum $\sum \Omega[L_i] . \Psi[R_i]$ over all 2-term factorizations of E, and this is a regular event. In a similar way, we see that $\Omega^{**}[E]$ is the sum of all expressions $\Omega[u]\Omega[v] \ldots \Omega[w]$ for which $uv \ldots w \in E$, counting the empty product when $1 \in E$, from which it follows that we have the matrix equality

$$\begin{bmatrix} \Omega^{**}[E_{11}] \ldots \\ \ldots \Omega^{**}[E_{nn}] \end{bmatrix} = \begin{bmatrix} \Omega[E_{11}] \ldots \\ \ldots \Omega[E_{nn}] \end{bmatrix}^* + \begin{bmatrix} o[E_{11}] \ldots \\ \ldots o[E_{nn}] \end{bmatrix}$$

where, of course, (E_{ij}) is the factor matrix of E. We might remark at this point that the output or constant part function o is none other than our biunit operator $[1, 1]$. The argument is more concisely expressed in terms of the factorial function of the previous chapter – if $E = E!(\ldots, E_k, \ldots)$ then we have $\Omega^{**}[E] = E!(\ldots, \Omega[E_k], \ldots)$. In future we shall discuss such results in these terms, and will also suppress all but the typical argument of a function of many variables, so that the above becomes simply 'if $E = E!(E_k)$ then $\Omega^{**}[E] = E!(\Omega[E_k])$'.

Corollary. *Every biregulator is a regulator.*

Proof. This follows from the Theorem and the observation that the operators $[a,1]$, $[1,a]$, $[b,1]$, $[1,b]$, ... are regulators.

It now follows from our table of biregulators that all the operators there defined are regulators for regular E, F, ... In particular, for instance, the shuffle of two regular events is regular.

The expansion and contraction operators

Let S be a free **S**-algebra, and introduce a new input letter x, so enlarging S to a new **S**-algebra $S\langle x \rangle = S_+$. We wish to consider two functions $x_+ : S \to S_+$ and $x_- : S_+ \to S$ called the *expansion* and *contraction* operators. If $E \in S$, then $x_+[E]$ is to be the sum of all events in S_+ of the form $x^*ux^*vx^* \dots x^*wx^*$ for which the word $uv \dots w \in E$. Again, if $E \in S_+$ we define $x_-[E]$ as the set of all words $uv \dots w \in S$ for which $x^*ux^*vx^* \dots x^*wx^*$ intersects E. These are not operators in the precise sense in which we have defined them, since they relate two distinct **S**-algebras, but we can extend them both to S_+-operators in a natural way, as we have done in the table, which shows that they are biregular and mutually dual.

Theorem 3. *Every biregulator on S is the restriction to S of a product of expansions, contractions, and regular intersections in a larger **S**-algebra.*

Proof. Let $\Omega = f((([a,1],[1,a],[b,1],[1,b],\dots))$, and introduce new letters $\bar{a}$, $\bar{b}$, ... Then if $E = f(\bar{a},a,\bar{b},b,\dots)$, the operator $\bar{\Omega} = a_-b_- \dots E.\bar{a}_+\bar{b}_+ \dots$ imitates Ω in the sense that if $\Omega[X] = g(a,b,\dots)$ then $\bar{\Omega}[X] = g(\bar{a},\bar{b},\dots)$. In verifying this, remember that we have $E^\cap = f(([\bar{a},\bar{a}],[a,a],[\bar{b},\bar{b}],[b,b],\dots))$. But for exactly the same reasons the image of $g(\bar{a},\bar{b},\dots)$ under the operator $\Psi = a_-b_- \dots F^\cap.a_+b_+ \dots$ where $F = (a\bar{a} + b\bar{b} + \dots)^*$ is $g(a,b,\dots)$, and so $\Psi\bar{\Omega}$ has the same effect as Ω on any member of S.

Corollary. *If* **X** *is any class of events (in differing free **S**-algebras) which is closed under expansions, contractions, and regular intersections, then* **X** *is closed under all biregulators, and in particular, under left and right multiplication and differentiation by regular events, shuffling with regular events, and so on. In a sense, the converse is true, since expansions* *and contractions and regular intersections are biregulators, but we must allow for the introduction of new letters if we are to have a formal theorem to this effect.*

The cross-product, and expansion and contraction functions

Let $f(x_1,\ldots,x_m)$ and $g(y_1,\ldots,y_n)$ be regular functions of the $m + n$ distinct input letters $x_1, \ldots, x_m, y_1, \ldots, y_n$. Then we define a new function $f \wedge g$ of mn variables z_{ij} ($1 \leqslant i \leqslant m$, $1 \leqslant j \leqslant n$) by the requirement that $z_{i_1 j_1} \cdots z_{i_k j_k} \in f \wedge g(z_{ij})$ if and only if we have both $x_{i_1} \cdots x_{i_k} \in f(x_i)$ and $y_{j_1} \cdots y_{j_k} \in g(y_j)$. (We are using the convention we threatened, whereby only a typical argument is indicated.) It is easy to see that $f \wedge g$ is regular, since

$$f \wedge g(z_{ij}) = f\left(\sum_j z_{ij}\right) \cap g\left(\sum_i z_{ij}\right).$$

The expansion function f_+ of $m + 1$ variables is defined by the equation $f_+(x_0, x_1, \ldots, x_m) = (x_0)_+[f(x_1, \ldots, x_m)]$, where x_0 is a new letter, and the contraction function f_- of $m - 1$ variables by

$$f_-(x_2, \ldots, x_m) = f(1, x_2, \ldots, x_m) = (x_1)_-[f(x_2, \ldots, x_m)].$$

It follows from these formulae that if **X** is any class of functions closed under intersection, multiplication by regular events, and the substitution of regular events for variables, then **X** is also closed under the formation of cross-product functions and expansion and contraction functions.

Theorem 4. *The product $\Omega\Psi$ of two biregulators is a biregulator.*

Proof. In the special case when $\Omega = f(([A_1, a_1], \ldots, [A_m, a_m]))$ and $\Psi = g(([b_1, B_1], \ldots, [b_n, B_n]))$ where the a_i and b_j are letters, not necessarily distinct, this is almost immediate, for then we have

$$[A_i, a_i] \cdot [b_j, B_j] = [0, 0] \text{ or } [A_i, B_j],$$

and so

$$\Omega\Psi = f(([A_i, a_i])) \cdot g(([b_j, B_j])) = f \wedge g(([A_i, a_i] \cdot [b_j, B_j]))$$

since a product of form $[A_{i_1} \ldots A_{i_h}, a_{i_1} \ldots a_{i_h}] \cdot [b_{j_1} \ldots b_{j_k}, B_{j_1} \ldots B_{j_k}]$ can be non-zero if and only if $h = k$ and $a_{i_p} = b_{i_p}$ for each p.

We reduce the general case to this by the judicious insertion of new letters. Let

$$\Omega = f(([a, 1], [1, a], [b, 1], [1, b], \ldots))$$

and

$$\Psi = g(([a, 1], [1, a], [b, 1], [1, b], \ldots)).$$

Then introduce the new letter x, and define two new operators Ω_+ and Ψ_+ by

$$\Omega_+ = f_+(([1, x], [a, x], [1, a], [b, x], [1, b], \ldots))$$

and

$$\Psi_+ = g_+(([x, 1], [a, 1], [x, a], [b, 1], [x, b], \ldots)).$$

Then a moment (or two) of thought convinces us that $\Omega\Psi = \Omega_+\Psi_+$, and this latter product has the special form we require. In the next chapter we shall prove a generalization of this theorem due to D. L. Pilling.

A dual pair of total regulators

The mutually dual operators $(I, 1)^*$ and $(1, I)^*$ are biregulators, since we can write them in the equivalent forms $(I, 1)^{**}$ and $(1, I)^{**}$. We follow Haines in proving the unexpected fact that these operators are both total regulators, i.e., take *any* events to regular events. The proof uses a little-known theorem of G. Higman which has other applications in this subject.

We introduce a notion of divisibility between words, writing $u|v$ to mean that $v \in (I, 1)^*[u]$, or equivalently that u can be obtained from v by deleting an arbitrary selection of the letters of v. Higman's theorem can then be stated in several equivalent ways:

Theorem 5. *Divisibility of words over a finite alphabet I has the properties*

(i) *In any set S of words the set of minimal words (those not divisible by another member of S) is finite.*

(ii) *In any infinite sequence $w_1, w_2, \ldots$ of words there is an infinite subsequence $w_i, w_j, w_k, \ldots$ with $w_i|w_j|w_k| \ldots$ and $i < j < k \ldots$.*

(iii) *There does not exist an infinite division-free sequence $w_1, w_2, \ldots$ (i.e., one for which $i < j$ implies $w_i \nmid w_j$).*

Proof. The equivalence of the three properties is fairly obvious, so we tackle only the form (iii). Supposing that division-free sequences exist, we select an 'earliest' one w_1, w_2, ... by the conditions:

w_1 is a shortest word beginning an infinite division-free sequence.
w_2 is a shortest word such that w_1, w_2 begins such a sequence.
w_3 is shortest such that w_1, w_2, w_3 begins such a sequence, and so on.

Then infinitely many of the w_i begin with the same letter, say $w_i = av_i$, $w_j = av_j$, $w_k = av_k$, ... for $i < j < k < \ldots$. Unfortunately, we then have the infinite division-free sequence $w_1, \ldots, w_{i-1}, v_i, v_j, v_k, \ldots$ which is 'earlier' than the one we chose.

As a corollary, we have Haines's theorems:

Theorem 6. *The operators* $(I,1)^*$ *and* $(1,I)^*$ *are total regulators.*

Proof. The event $(I,1)^*[E]$ is the set of all words divisible by some word of E. It is obviously equal to $(I,1)^*[F]$, where F is the set of minimal words of E, and since F is finite this is regular. The event $G = (1,I)^*[E]$ is the set of all words dividing some word of E. We prove that G has only finitely many distinct word derivates $\partial G/\partial w$. If not, we can by Higman's theorem produce an infinite sequence of distinct derivates $\partial G/\partial w_1$, $\partial G/\partial w_2$, ... with $w_1|w_2|w_3|$ .... It follows that we have

$$\frac{\partial G}{\partial w_1} \supset \frac{\partial G}{\partial w_2} \supset \frac{\partial G}{\partial w_3} \supset \ldots,$$

and so we can select words

$$v_1 \in \frac{\partial G}{\partial w_1} \Big\backslash \frac{\partial G}{\partial w_2}, v_2 \in \frac{\partial G}{\partial w_2} \Big\backslash \frac{\partial G}{\partial w_3}, \ldots.$$

But then the sequence $v_1, v_2, v_3, \ldots$ is an infinite division-free sequence, contradicting Higman's theorem.

By using other divisibility relations with similar properties we can generalize these results. Pilling proves in this way that for any operator Ω the operator $\{(1,I)^*:\Omega\}^*$ and its dual are both total regulators.

CHAPTER EIGHT

Event classes and operator classes

Let E be any event in a free **S**-algebra over a finite input alphabet, $I = \{a_1, a_2, \ldots\}$. We can convert E into a function defined in *all* **S**-algebras by defining $E(A_1, A_2, \ldots)$ to be the sum of all products $A_{i_1} A_{i_2} \ldots A_{i_n}$ (as n varies) for which $a_{i_1} a_{i_2} \ldots a_{i_n} \in E$. Thus for instance to the event $E = \{a^n b^n | n \geqslant 0\}$ there corresponds the function $E(A, B) = \sum A^n B^n$ (over $n \geqslant 0$). The functions that arise in this way we call **S**-functions. Of course not every function defined in all **S**-algebras is an **S**-function – we instance the function $A \cap B$ defined as the sum of all X for which $X \leqslant A$, $X \leqslant B$.

More generally, if **X** is any class of events (possibly in different free **S**-algebras), we obtain in this way from the events of **X** the class of **X**-*functions*. We shall use **S** for the class of all events, **R** for the class of all regular events, and **P** for the class of all finite events (or *polynomials*). Thus the **R**-functions are just the regular functions, and the **P**-functions in variables $A_1, \ldots, A_k$ are just the finite sums of products of the A_i (including possibly the empty product 1). We shall not usually distinguish between the class **X** of events and the corresponding class of functions – thus we write $f \in \mathbf{X}$ to mean that f is an **X**-function, and speak simply of *classes*. In view of this convention, we had better confine ourselves to classes which are invariant under renaming of letters in any 1–1 fashion.

If **F**, **A**, **B**, . . . are classes, then $\mathbf{F}(\mathbf{A}, \mathbf{B}, \ldots)$ denotes the class of all events (and the corresponding functions) of the form $f(X_1, X_2, \ldots)$, where each X is in one of **A**, **B**, . . ., and $f \in \mathbf{F}$. Usually if $\varphi(A, B, \ldots)$

is any notion defined for events A, B, ... we define

$$\varphi(\mathbf{A}, \mathbf{B}, \ldots) = \{\varphi(A, B, \ldots) | A \in \mathbf{A}, B \in \mathbf{B}, \ldots\}.$$

There is one important exception – $A \cap \mathbf{B}$ will denote the intersection of the classes $\mathbf{A}$ and $\mathbf{B}$ – we can use $\mathbf{A}^{\cap}[\mathbf{B}]$ or $\mathbf{A}^{\cap}\mathbf{B}$ for $\{A \cap B | A \in \mathbf{A}, B \in \mathbf{B}\}$. We shall also abbreviate $\mathbf{C}(([\mathbf{A}, \mathbf{B}]))$ to $[\mathbf{A}, \mathbf{B}]_{\mathbf{C}}$ – this is the class of all operators of the form $c(([A_1, B_1], \ldots, [A_k, B_k]))$ for which $c \in \mathbf{C}$, $A_i \in \mathbf{A}$, $B_i \in \mathbf{B}$. We use $\langle \mathbf{A}, \mathbf{B} \rangle$ for the class of all operators Ω for which $B \in \mathbf{B}$ implies $\Omega[B] \in \mathbf{A}$. Thus $\langle \mathbf{R}, \mathbf{R} \rangle$ is the class of all regulators, and $\langle \mathbf{R}, \mathbf{S} \rangle$ the class of total regulators.

Theorem 2 of the last chapter generalizes immediately to yield

Theorem 1. $\mathbf{R}((\langle \mathbf{X}, \mathbf{R} \rangle)) \subseteq \langle \mathbf{R}(\mathbf{X}), \mathbf{R} \rangle$.

Corollary. $[\mathbf{X}, \mathbf{S}]_{\mathbf{R}} \subseteq \langle \mathbf{R}(\mathbf{X}), \mathbf{R} \rangle$.

In particular, operators of any class $[\mathbf{R}, \mathbf{X}]_{\mathbf{R}}$ are regulators. We call such operators $\mathbf{X}$-*regulators*.

Theorem 2. $[\mathbf{S}, \mathbf{X}]_{\mathbf{Y}} \cap \langle \mathbf{P}, \mathbf{P} \rangle \subseteq [\mathbf{P}, \mathbf{X}]_{\mathbf{Y}}$.

Proof. We consider $\Omega = f(([S_1, X_1], \ldots, [S_n, X_n]))$ with $f \in \mathbf{Y}$, $X_i \in \mathbf{X}$. Then Ω is a sum of terms of the form $[S_i S_j \ldots S_k, X_i X_j \ldots X_k]$. Now, if any one of the events $S_i, S_j, \ldots, S_k, X_i, X_j, \ldots, X_k$ in such a term is zero the term reduces to the zero operator, and we call if a *null term*. Any of the events $S_1, \ldots, S_n$ which appears only in null terms, or does not appear at all, may be replaced by 0 without affecting Ω. The others must all be finite, for if $[S_i S_j \ldots S_k, X_i X_j \ldots X_k]$ is a non-null term and $x_i \in X_i, \ldots, x_k \in X_k$, then $S_i S_j \ldots S_k \leqslant \Omega[x_i x_j \ldots x_k]$ and so is a finite event.

In particular, if a member of $[\mathbf{S}, \mathbf{S}]_{\mathbf{R}}$ preserves finiteness, it also preserves regularity (since $[\mathbf{S}, \mathbf{S}]_{\mathbf{R}} \cap \langle \mathbf{P}, \mathbf{P} \rangle \subseteq [\mathbf{P}, \mathbf{S}]_{\mathbf{R}}$). The same argument proves that any $\Omega \in [\mathbf{S}, \mathbf{S}]_{\mathbf{X}}$ such that both Ω and $^\partial\Omega$ preserve finiteness is in fact in $[\mathbf{P}, \mathbf{P}]_{\mathbf{X}}$. In particular, if Ω and $^\partial\Omega$ preserve finiteness and $\Omega \in [\mathbf{S}, \mathbf{S}]_{\mathbf{R}}$, then Ω is a biregulator.

We can now prove that neither of the operators $\leftarrow$, $\bigcirc$ of the last chapter is in $[\mathbf{S}, \mathbf{S}]_{\mathbf{R}}$. If so, since both are self-dual and preserve finiteness, they would be biregulators. Let a, b be distinct letters of the alphabet. Then

$$(a^* b^*)^\cap . \leftarrow . (b^* a^*)^\cap = (a^* b^*)^\cap . \bigcirc . (b^* a^*)^\cap = \sum [a^m b^n, b^n a^m]$$

(over all m, n) $= \Omega$, say. It suffices therefore to prove that Ω is not a biregulator. Suppose that u, v, w, x, y, z are words such that

$$[u,v]:[w,x]^{**}:[y,z] \leqslant \Omega.$$

Then for each $p \geqslant 0$ there are m, $n \geqslant 0$ such that $uw^p y = a^m b^n$ and $vx^p z = b^n a^m$. It follows first that w and x are of the same length, second that neither involves both a and b, and finally that $w = x$.

From this it follows that whenever Ψ^{**} appears in the expression for Ω as a biregulator, then $\Psi = E^\cap$ for some E, and so $\Psi^{**} = (E^*)^\cap$. We therefore deduce that Ω is a finite sum of biproducts of terms of the form $[u,v]$ or $E^\cap$, and this is easily disproved.

We have skated past what we might call the 'u, v, w, x, y, z' lemma for members Ω of $[\mathbf{S}, \mathbf{S}]_\mathbf{R}$. If such an operator is not in $[\mathbf{S}, \mathbf{S}]_\mathbf{P}$ there must be words u, v, w, x, y, z with $wx \neq 1$ for which $[uw^p y, vx^p z] \in \Omega$ for all p. There is a corresponding 'u, v, w lemma' for infinite regular events E – there are words u, v, w with $v \neq 1$ and $uv^p w \in E$ for all p. In Chapter 10 we shall see that there is a similar 'u, v, w, x, y lemma' for infinite context-free languages.

Homomorphisms. The non-zero operator Ω is called a *homomorphism* if it satisfies $\Omega[XY] = \Omega[X]\Omega[Y]$ for all X, Y, and a *unit-homomorphism* if also $\Omega[1] = 1$. A homomorphism is called an **X**-*homomorphism* if its values at 1 and the inputs are members of the class **X**. It is a *semigroup homomorphism* if its values at words are words. Most authors restrict the term 'homomorphism' to this special case, and use 'substitution' for our 'unit-homomorphism'.

Theorem 3. *Let $\Omega[1] = U$, $\Omega[a] = A$, $\Omega[b] = B$, ... Then Ω is a homomorphism if and only if $\Omega = ([U, 1] + [A, a] + [B, b] + \ldots)^{**}$, and is a unit-homomorphism if and only if $\Omega = ([A, a] + [B, b] + \ldots)^{**}$. Every* **X**-*homomorphism is a member of* $[\mathbf{X}, \mathbf{R}]_\mathbf{R}$, *and so its dual is an* **X**-*regulator. An* **R**-*homomorphism is a biregulator.*

Proof. If Ω is a homomorphism, we must have $U^* = U$ and $UAU = A$, etc., since $1^* = 1$, and $1a1 = a$, etc. We also have

$$\Omega[a_1 a_2 \ldots a_k] = A_1 A_2 \ldots A_k,$$

where $A_i = \Omega[a_i]$, and from these equations we get

$$\Omega = ([U,1] + [A,a] + [B,b] + \ldots)^{**}.$$

The converse implication follows from the observation that any operator of the form $([U,1] + [A,a] + \ldots)^{**}$ is a homomorphism, and the statements about unit-homomorphism are similarly proved. The remaining statements are obvious consequences of earlier theorems.

Corollary. *Any operator of the form* $((1,U) + (a,A) + \ldots)^*$ *(note the round brackets and the single star) is a regulator.*

Proof. If we think carefully about the action of the dual operator $\Omega = ((U,1) + (A,a) + \ldots)^*$ we conclude rapidly that $\Omega[uv] = \Omega[u]\Omega[v]$, so that Ω is a homomorphism.

The structure of the operators [A, B]_C

Let $\Omega = f(([A_1,B_1],\ldots,[A_n,B_n])), f \in \mathbf{C}, A_i \in \mathbf{A}, B_i \in \mathbf{B}$, and let $c_1, \ldots, c_n$ be distinct new letters. Then in the extended alphabet we can factorize Ω as

$$\Omega = ([A_1,c_1] + \ldots + [A_n,c_n])^{**}.\Psi.([c_1,B_1] + \ldots + [c_n,B_n])^{**},$$

or say $\Omega = \alpha.\Psi.\beta$, where $\Psi = f(([c_1,c_1],\ldots,[c_n,c_n]))$. But plainly $\Psi = C^\cap$, where C is the $\mathbf{C}$-event $f(c_1,\ldots,c_n)$, and the operator α is a unit $\mathbf{A}$-homomorphism, while β is the dual of a unit $\mathbf{B}$-homomorphism.

We conclude

Theorem 4. *Any operator* $\Omega \in [\mathbf{A},\mathbf{B}]_\mathbf{C}$ *may be factorized into the product of a unit* $\mathbf{A}$-*homomorphism, an intersection with a* $\mathbf{C}$-*event, and the dual of a unit* $\mathbf{B}$-*homomorphism.*

(We have described the product in the order of writing, which is of course the opposite of the order of application.) We shall not use this result, but it is obviously a sensible way to think of this kind of operator.

The Pilling product theorem

We ask what can be said about the product of operators from the two classes $[\mathbf{A},\mathbf{B}]_\mathbf{C}$ and $[\mathbf{D},\mathbf{E}]_\mathbf{F}$? Does it necessarily belong to some class

[G, H]$_I$ where **G, H, I** are classes depending on **A, B, C, D, E, F**? Rather unexpectedly, it seems that useful results of this type exist only when **B** = **C** = **R** or **D** = **F** = **R**. The two results are mutually dual (so that either is asymmetric) and we can state the basic theorem in the form

Theorem 5. [**A, R**]$_R$. [**D, E**]$_F$ ⊆ [(**R** ∧ **D**$_+$)(**A**), **E**]$_{RVF}$, *provided* 1 ∈ **A**.

Proof. We first observe that since $[A, B] = [A, 1] . [1, B]$ and since for regular $B [1, B]$ is a biregular function of the operators $[1, a], [1, b], \ldots,$ we can write any operator of [**A, R**]$_R$ in a form $c(([A_i, b_i]))$ in which c is regular, $A_i \in$ **A**, each b_i is 1 or a letter, and we indicate only the typical argument of c. We now write down a series of equations which prove the theorem, and follow with their explanation.

$$c(([A_i, b_i])) . f(([D_j, E_j])) \tag{1}$$
$$= c!((c_k(([A_i, b_i])))) . f(([D_j, E_j])) \tag{2}$$
$$= (c! \wedge f)((c_k(([A_i, b_i])) . [D_j, E_j])) \tag{3}$$
$$= (c! \wedge f)((c_k((A_i, b_{i+}))] . [D_{j+}, E_j])) \tag{4}$$
$$= (c! \wedge f)(([(c_k \wedge D_{j+})(A_i), E_j])) \tag{5}$$

In (1), we suppose the b_i are 1 or letters, $c \in$ **C**, $A_i \in$ **A**, $D_j \in$ **D**, $E_j \in$ **E**, $f \in$ **F**. Line (2) is obtained by expressing the function c in terms of its factorial function and factor functions, and the equality with line (3) follows from factor theory. In line (4) we have introduced a new letter x, defining $b_{i+} = x$ when $b_i = 1$, and $b_{i+} = b_i$ otherwise, and $D_{j+} = x_+[D_j]$ to compensate. To get line (5) we regard D_{j+} as a function of the letters $x, a, b, \ldots$ of the enlarged alphabet, take all words in these letters which belong both to $c_k(b_{i+})$ and to D_{j+}, and then allow ourselves to replace any letter b_{i+} by any corresponding letter A_i. In the resulting formula (5) the events and functions belong to the appropriate classes.

Corollaries. (i) [**E, D**]$_F$. [**R, A**]$_R$ ⊆ [**E**, (**D**$_+$ ∧ **R**)(**A**)]$_{F\wedge R}$.

(ii) [**A, R**]$_R$ [**D**] ⊆ (**R** ∧ **D**$_+$)(**A**).

(iii) [**E, D**]$_F$ [**R**] ⊆ (**F** ∧ **R**)(**E**).

Proofs. (i) is just the dual of the theorem. To prove (ii), observe that $\Omega . [D, 1] = [\Omega[D], 1]$, take $E = 1$, f the identity function, and re-interpret line (5) of the proof, (iii) is similar from (i).

Now, let L be any class such that $L_+ = L \wedge R = L(L) = L$. Then we have

Theorem 6. $[L, R]_R \cdot [L, R]_R \subseteq [L, R]_R \subseteq \langle L, L \rangle$

$\qquad\quad [R, L]_R \cdot [R, L]_R \subseteq [R, L]_R \subseteq \langle R, R \rangle$

$\qquad\quad [L, S]_L \subseteq \langle L, R \rangle$.

Proof. These are immediate from Theorem 5 and its corollaries.

In Chapter 10 we shall specialize the class L to be the class of context-free languages, which does indeed have these properties.

Ginsburg and Greibach call a non-empty class a full AFL (Abstract Family of Languages) if it is closed under the regular operations $+, ., *$, under regular intersections, and regular homomorphisms, and each event of the class involves only finitely many letters. (They also have a concept of 'unfull' AFL, which we find less interesting.) We can easily deduce from our theorems that

Theorem 7. *A class* X *is a full AFL if and only if* $R(X) = X$ *and* $[R, R]_R [X] = X$, *that is to say, if and only if it is closed under regular and biregular functions. For any class* X, *the class* $R([R, R]_R [X])$ *is a full AFL. If* X *and* Y *are full AFL's, so is* $X(Y)$.

Proof. Since expansions and contractions are regular homomorphisms, full AFL's are closed under them, and therefore under biregulators. A class closed under biregulators and regular functions is *a fortiori* closed under regular intersections and regular homomorphisms. The last statement is proved using the formula

$$E \cap f(\ldots, A_j, \ldots) = (E ! \wedge f)(\ldots, E_k \cap A_j, \ldots)$$

for a regular event $E = E !(\ldots, E_k, \ldots)$.

CHAPTER NINE

Some regulator algebras

Definition. A *regulator algebra* is any subset of $\langle \mathbf{R}, \mathbf{R} \rangle$ closed under the regular operations $+$, $.$, $*$.

The discovery of interesting regulator algebras is a significant part of the study of closure properties of the regular events.

Theorem 1. *Let R be a regulator algebra. Then the class of all operators of the form $\Omega + \Psi$ ($\Omega \in R$, $\Psi \in \langle \mathbf{R}, \mathbf{S} \rangle$) is a regulator algebra. In other words, any regulator algebra can be enlarged to a regulator algebra which contains all total regulators.*

Proof. We have

$$(\Omega + \Psi) + (\Omega' + \Psi') = (\Omega + \Omega') + (\Psi + \Psi'),$$

$$(\Omega + \Psi).(\Omega' + \Psi') = (\Omega\Omega') + (\Omega\Psi' + \Psi\Omega' + \Psi\Psi'),$$

and

$$(\Omega + \Psi)^* = \Omega^* + \Omega^*.\Psi.(\Omega + \Psi)^*.$$

bearing in mind that if Ψ is a total regulator it is any operator $\Psi\Omega$, while $\Omega\Psi$ is a total regulator if Ω is a regulator, we see that these formulae enable us to express the sum, product, and star of operators of the form $\Omega + \Psi$ ($\Omega \in R$, $\Psi \in \langle \mathbf{R}, \mathbf{S} \rangle$) in the same form.

Corollary. *Any regulator algebra may be enlarged so as to contain all operators of the form E^+ ($E \in \mathbf{R}$).*

Proof. This follows from the equation $E^+ = 1 + [E, I^*]$, the sum of the identity operator and a total regulator.

Pilling conjectures that a corresponding result holds with the regular additions E^+ replaced by the regular intersections E, i.e., that any regulator algebra may be enlarged so as to contain all regular intersections. The conjecture is probably false, at least in this generality, but it certainly holds for the most interesting regulator algebras.

Definition. The operator Ω is *open* if $\Omega[EFG] \geqslant E\Omega[F]G$ for all E, F, G. It is *convex* if $\Omega[EF] \leqslant \Omega[E]\Omega[F]$, and *concave* if $\Omega[EF] \geqslant \Omega[E]\Omega[F]$ for all E, F.

Theorem 2. Ω *is open if and only if* $\Omega[EF] \geqslant \Omega[E]F + E\Omega[F]$. *The star of a convex operator is convex, the star of an open operator is open and concave, and the star of a convex open operator is a homomorphism.*

Proof. If Ω is open then $\Omega[EF] = \Omega[EF1] \geqslant E\Omega[F]1$, and similarly $\Omega[EF] \geqslant \Omega[E]F$, so that $\Omega[EF] \geqslant \Omega[E]F + E\Omega[F]$. But if Ω satisfies this inequality then $\Omega[EFG] \geqslant E\Omega[FG] \geqslant E\Omega[F]G$, so that Ω is open. The fact that if Ω is convex or open so is Ω^*, is almost obvious. Now, if Ω is open, so is Ω^n, by induction, and so

$$\Omega^*[EF] \geqslant \Omega^{m+n}[EF] \geqslant \Omega^m[E]\Omega^n[F],$$

from which we get $\Omega^*[EF] \geqslant \Omega^*[E]\Omega^*[F]$ by summing over all m, n. If Ω is also convex, then Ω^* will be both concave and convex, and so a homomorphism.

Theorem 3. *Each of the following sets*

  open dual-convex regulators

  open dual-convex **S**-regulators

  finite sums of the duals of open homomorphisms

is a regulator algebra. (Ω *is dual-convex if* $^\partial\Omega$ *is convex.*)

Proof. The closure under sum and product is obvious in every case. Now if Ω belongs to any one of the three sets, then $^\partial\Omega$ is open and convex, so that $(^\partial\Omega)^*$ is a homomorphism, whence Ω^* belongs to $[\mathbf{R}, \mathbf{S}]_\mathbf{R}$ by Theorem 3 of Chapter 8, so that Ω^*, being open, belongs to all of the sets.

In the next chapter we shall extend this slightly so as to prove that the open dual-convex **L**-regulators from a regulator algebra, **L** being the class of context-free languages.

A more typical proof that the operators of a certain kind form a regulator algebra has the following shape. One expresses any regular function of the generating operators in a 'normal form' in which the generators appear in such a simple way that it is obvious that the operator in normal form is a regulator. Thus in Theorem 1 we found the normal form $\Omega + \Psi$ ($\Omega \in R$, $\Psi \in \langle \mathbf{R}, \mathbf{S} \rangle$) for the regulator algebra generated by members of R and total regulators. In more complicated cases it might happen that the passage to the normal form is accomplished by applying a certain regulator to the regular function involved, so that theorems of this type can be self-generating.

We now consider regular functions of operators of the form E^l, ∂E^l, $_\partial E^l$, E^r, ∂E^r, $_\partial E^r$, E^+, $E^\cap$ for regular events E.

Theorem 4. $\mathbf{R}(\mathbf{R}^l, \mathbf{R}^r) \nsubseteq \langle \mathbf{R}, \mathbf{R} \rangle$. *On the other hand,*

$$\mathbf{R}((\mathbf{R} + 1)^l, (\mathbf{R} + 1)^r) \subseteq \langle \mathbf{R}, \mathbf{R} \rangle.$$

Since of course $\mathbf{R} + 1$ is the class of all regular events $\geqslant 1$, the theorem asserts that operators $f(E^l, E^r, F^l, F^r, \ldots)$ need not be regulators, but that they are regulators if $E \geqslant 1$, $F \geqslant 1$, $\ldots$.

Proof. The operator $(a^l b^r)^*$ takes 1 to the irregular event $\sum a^n b^n$, summed over all n, proving the first part. Now we prove the second part by showing that each operator in $\mathbf{R}((1 + \mathbf{R})^l, (1 + \mathbf{R})^r)$ can be expressed as a finite sum of terms $E^l E^r$ ($E, F \in 1 + \mathbf{R}$). This is obvious for sums, and for products since $E^l F^r . G^l H^r = (EG)^l (HF)^r$. Now the star of $E^l F^r$ is $\sum (E^n)^l (F^n)^r$ over all n, which is the same as $\sum (E^m)^l (F^n)^r$ over all m, n, since $E \geqslant 1$, $F \geqslant 1$, and this second sum is $(E^*)^l (F^*)^r$. Then $(E_1^l F_1^r + \ldots + E_k^l F_k^r)^* = (E_1 E_2 \ldots E_k)^{*l} (F_k \ldots F_2 F_1)^{*r}$, since for $A \geqslant 1$, $B \geqslant 1$, $\ldots$ we have $(A + B + \ldots)^* = (AB \ldots)^*$.

For more complicated algebras the production of the normal form requires more technique, so we now prove some theorems on the conversion of expressions to normal form in appropriate circumstances.

Theorem 5. *Let S be an **S**-algebra, containing an **R**-algebra R, and a finite number of elements $f_1, \ldots, f_n$. If R is generated by elements*

$r_1, r_2, \ldots$ *and every element of the form* $r_i f_j$ *or* $f_i f_j$ *or* 1 *can be put into the form* $f_1 R_1 + \ldots + f_n R_n$ $(R_i \in R)$, *then the set* R' *of elements so expressible is an* **R**-*algebra*.

Proof. Let g be any one of $f_1, \ldots, f_n, r_1, r_2, \ldots$. Then there exist elements G_{ij} of R such that $gf_j = \sum f_i G_{ij}$ (summed over i). Now represent the element $x = \sum f_j X_j$ of R' by the vector X whose components are the X_j. Then the effect of premultiplying x by g is to premultiply X by the matrix G whose components are the G_{ij}. In this way every generator g of R' corresponds to a matrix G, and premultiplying x by $f(g_1, g_2, \ldots)$ corresponds to premultiplying X by $f(G_1, G_2, \ldots)$. Since $1 \in R'$, we get $f(g_1, g_2, \ldots) \in R'$, by premultiplying!

Theorem 6. *Let* S *be an* **S**-*algebra, with regular subalgebras* L *and* R *generated by elements* $l_1, l_2, \ldots$ *and* $r_1, r_2, \ldots$ *respectively. Suppose that every product* $r_i l_j$ *can be expressed as a finite sum* σ_{ij} *of terms of the form* l_k, r_k, *or* 1. *Then the set of finite sums of terms of the form* $L_i R_i$ $(L_i \in L, R_i \in R)$ *is an* **R**-*algebra*.

Proof. If in any product π of factors of the form l_k or r_k a factor r_i immediately precedes a factor l_j we can express π as a sum of simpler products on replacing $r_i l_j$ by σ_{ij}. Continuing in this way we eventually express π as a sum of products in no one of which does an r_i precede an l_j.

Now regard the symbols $l_1, \ldots, r_1, \ldots$ as distinct input letters in a *free* **S**-algebra, and let Ω be the operator $\sum (\sigma_{ij}, r_i l_j)$ summed over all pairs i, j. Then the above reduction process applied to any **S**-function $f(l_1, \ldots, r_1, \ldots)$ becomes:

    (i) regard l_i, r_j as distinct input letters in a free **S**-algebra;
    (ii) apply the operator Ω^*;
    (iii) intersect with the regular event $(l_1 + \ldots)^* (r_1 + \ldots)^* = E$, say;
    (iv) restore the original meanings for l_i, r_j.

Since Ω^* is a regulator, by the Corollary to Theorem 3 of the last chapter, for a *regular* function f we obtain after stage (ii) a regular event, and since E is regular, after stage (iii) we have an event which can be expressed by the last theorem of Chapter 6 as a finite sum of terms $g(l_1, \ldots) h(r_1, \ldots)$ in which g and h are regular, so that after stage (iv) we have the desired expression for $f(l_1,\ldots,r_1,\ldots)$ as a finite sum of products $L_i R_i$ ($L_i \in L$, $R_i \in R$).

Theorem 7. *Let S be an **S**-algebra, in which we have regular subalgebras L and R and a finite subalgebra $F = \{f_1,\ldots,f_n\}$. Suppose further that (l_i, r_j being generators for L and R respectively):*

   (i) *each product $r_i f_j$ is a finite sum of terms $f_s R_t$ ($R_t \in R$)*
   (ii) *each product $f_i l_j$ is a finite sum of terms $L_s f_t$ ($L_s \in L$)*
   (iii) *each product $r_i f_j l_k$ is a finite sum of terms $f_s r_t f_u$ or $f_s l_t f_u$.*

*Then the set of finite sums of the form $L_i f_j R_k$ ($L_i \in L$, $R_k \in R$) is an **R**-algebra.*

Proof. We can suppose that 1 appears among the r_i, also among the l_j. Let L' be the **R**-algebra generated by all products $f_i l_j f_k$, and R' that generated by all products $f_i r_j f_k$. Then the hypotheses of this theorem imply that those of the last theorem hold for L' and R', so that any regular function of the l_i, f_j, r_k can be expressed as a finite sum of terms $L'_i R'_i$ ($L'_i \in L'$, $R'_i \in R'$). But then Theorem 5 implies that each member of R' can be written as a finite sum of terms $f_i R_i$ ($R_i \in R$), and this together with the corresponding statement about L and F establishes the theorem.

These theorems have several applications, but we give only one.

Theorem 8. $\mathbf{R}(\mathbf{R}^l, {}^{\partial}\mathbf{R}, \mathbf{R}^{\cap})$ *is a regulator algebra. Indeed, every regular function of operators of the form E^l, $F^{\cap}$, ${}^{\partial}G$ (E, F, $G \in \mathbf{R}$) can be expressed in normal form as a finite sum of terms $E^l . F^{\cap} . {}^{\partial}G$ ($E, F, G \in \mathbf{R}$).*

Proof. The typical operator of this type can be written as

$$f(a^l, b^l, \ldots, {}^{\partial}a, {}^{\partial}b, \ldots, E_1^{\cap}, \ldots, E_n^{\cap}),$$

where $a, b, \ldots$ are letters and $E_1, \ldots, E_n$ regular events. Now the class C of all Boolean functions of word derivates of $E_1, \ldots, E_n$ is a finite class closed under Boolean functions and word derivation, and if E, F belong to this class, we have

$$E^{\cap} + F^{\cap} = (E + F)^{\cap}, \qquad E^{\cap} . F^{\cap} = (E \cap F)^{\cap},$$

$$(E^{\cap})^* = 1 = (I^*)^{\cap},$$

so that the operators $E^\cap$ ($E \in C$) form a finite regulator algebra. Using the identities

$$^\partial a . E^\cap = \left(\frac{\partial E}{\partial a}\right)^\cap . {}^\partial a, \qquad E^\cap . a^l = a^l . \left(\frac{\partial E}{\partial a}\right)^\cap ,$$

and $^\partial a . a^l = 1$, $^\partial a . b^l = 0$ ($a \neq b$) we can reduce any expression $^\partial a . E^\cap . b^l$ ($E \in C$) to a form $F^\cap$ ($F \in C$), and so Theorem 7 applies.

The above proof is rather extravagant. If Ω involves three intersection operators $E^\cap$, $F^\cap$, $G^\cap$, and E, F, G have three word derivates each, then the proof as given will in some cases involve taking the star of a $2^{512} \times 2^{512}$ matrix! We can reduce this to a 512×512 matrix by noting that the f_i in Theorem 7 need not be all the elements of F, but merely an additive basis (so that every member of F can be expressed as a sum of f_i). Readers who have tried to find the star of a 4×4 matrix will appreciate that there will still be problems! I think the problems are genuine, in that there will be cases of this size for which the normal form really requires 512 terms.

If Ω involves only operators of two of the types E^l, $F^\cap$, $^\partial G$, then the same will be true of the normal form, whose derivation can be simplified. Thus in the cases $^\partial E$, $E^\cap$ and E^l, $E^\cap$ we can use Theorem 5, and in the case E^l, $^\partial E$ Theorem 6, with the l_i, r_j being the operators a^l, $^\partial b$. In particular cases there will be *ad hoc* techniques, and it is seldom wise to employ the general argument.

Theorem 9. $R(^\partial R, {}^\partial R^r, R^\cap)$ *is a regulator algebra.*

Proof. We prove a more general theorem. If

$$\Omega = f(E_1^\cap, \ldots, E_n^\cap, {}^\partial a, a^r, \ldots)$$

is any **S**-function of its arguments, then Ω is a regulator. We can suppose that the set $\{E_1, \ldots, E_n\}$ is closed under left and right word differentiation and Boolean functions. Then, since $^\partial a . E^\cap = (\partial E / \partial a)^\cap . {}^\partial a$ and $^\partial a^r . E^\cap = (\partial E / \partial a^r)^\cap . {}^\partial a^r$, $E^\cap . F^\cap = (E \cap F)^\cap$, we can reduce any product of terms $E_i^\cap$, $^\partial a$, $^\partial a^r$ to a form $E_j^\cap . \pi$, where π is a product involving no term $E_k^\cap$. It follows that $\Omega[X]$ is a sum of terms $E_j \cap \pi[X]$, and since for a regular X, $\pi[X]$ takes only finitely many values, this is a regular expression for $\Omega[X]$.

Theorem 10. *The algebras*

$$R(R^l, {}^\partial R^l, R^\cap, R^+, {}_\partial R, {}_\partial R^r), \qquad R(R^r, {}^\partial R^r, R^\cap, R^+, {}_\partial R, {}_\partial R^r),$$

and

$$\mathbf{R}(^{\partial}\mathbf{R}^l, {}^{\partial}\mathbf{R}^r, \mathbf{R}^{\cap}, \mathbf{R}^+, {}_{\partial}\mathbf{R}^l, {}_{\partial}\mathbf{R}^r)$$

are regulator algebras.

Proof. We enlarge the regulator algebras

$$\mathbf{R}(\mathbf{R}^l, {}^{\partial}\mathbf{R}^l, \mathbf{R}^{\cap}), \qquad \mathbf{R}(\mathbf{R}^r, {}^{\partial}\mathbf{R}^r, \mathbf{R}^{\cap}) \qquad \text{and} \qquad \mathbf{R}(^{\partial}\mathbf{R}^l, {}^{\partial}\mathbf{R}^r, \mathbf{R}^{\cap})$$

given by Theorems 8 and 9 so as to contain all total regulators, and so all the operators of $\mathbf{R}^+$. The operators of ${}_{\partial}\mathbf{R}$ and ${}_{\partial}\mathbf{R}^r$ were proved to be total regulators in Theorem 3 of Chapter 5.

On the other hand, we have already shown that $\mathbf{R}^l$ and $\mathbf{R}^r$ cannot simultaneously be part of a regulator algebra, since $(a^l b^r)^*$ is not a regulator, and the observation that $(^{\partial}a \cdot a^r + {}^{\partial}b \cdot b^{2r})^*$ is not a regulator shows that $^{\partial}\mathbf{R}$ and $\mathbf{R}^r$ (and similarly $^{\partial}\mathbf{R}^r$ and $\mathbf{R}^l$) cannot both be included in a regulator algebra. (Let $\Omega = a \cdot a^r + b \cdot b^{2r}$. Then under the action of Ω we have:

$$ab \to ba \to ab^2 \to b^2 a \to bab^2 \to ab^4 \to \ldots \to ab^8 \to \ldots \to ab^{16} \to \ldots,$$

and so we deduce $^{\partial}a \cdot \Omega^*[ab] = b + b^2 + b^4 + b^8 + \ldots$, an irregular event.) In fact we shall prove in Chapter 15 that operators of type $\mathbf{R}(^{\partial}\mathbf{R}, \mathbf{R}^r)$ can be used to imitate the general Turing machine.

So if $\mathbf{U}$ is some union of the operator classes $\mathbf{R}$, $^{\partial}\mathbf{R}$, ${}_{\partial}\mathbf{R}$, $\mathbf{R}^r$, $^{\partial}\mathbf{R}^r$, ${}_{\partial}\mathbf{R}^r$, $\mathbf{R}^+$, $\mathbf{R}^{\cap}$, then $\mathbf{R}(\mathbf{U})$ is a regulator algebra if and only if $\mathbf{U}$ contains no one of $\mathbf{R}^l \cup {}^{\partial}\mathbf{R}^r$, $^{\partial}\mathbf{R}^l \cup \mathbf{R}^r$, and $\mathbf{R}^l \cup \mathbf{R}^r$.

In Theorem 10, we can replace either of the classes $^{\partial}\mathbf{R}$, $^{\partial}\mathbf{R}^r$ by the corresponding class $^{\partial}\mathbf{S}$, $^{\partial}\mathbf{S}^r$. For the third of the regulator algebras there given, this follows from the proof of Theorem 9. But for the first and second we seem to need a new approach, since there does not seem to be a normal form for such operators. We compress the proof, since it does not seem particularly important, but perhaps the method will have other applications.

Theorem 11. $\mathbf{R}(\mathbf{R}^r, {}^{\partial}\mathbf{S}^r, \mathbf{R}^{\cap}) \subseteq \langle \mathbf{R}, \mathbf{R} \rangle$.

Proof. We let $\Omega = f(\Omega_1, \ldots, \Omega_n)$ with each $\Omega_i \in \mathbf{R}^r \cup {}^{\partial}\mathbf{S}^r \cup \mathbf{R}^{\cap}$, and make use of the matrix normal form (Chapter 3, Theorems 7 and 8) for the regular function f. Thus we write $\Omega = LM^*N$, where L and N are row and column vectors with entries 0 and 1, and M is a matrix whose typical entry M_{ij} is $E_{ii}^r + \partial F_{ij}^r + G_{ij}^{\cup}$, a sum of the operators Ω_i. Now let X, Y be the column vectors with $X_j = N_j Z$, $Y = M^*[X]$ for some regular event Z. Then, since $\Omega[Z] = LY$ we need only prove the Y_i regular. But we have

$$Y_i = X_i + \sum \left(Y_j E_{ij} + \frac{\partial Y_j}{\partial F_{ij}^r} + G_{ij} \cap Y_j \right) \text{ (summed over } j)$$

and so for any word w

$$\frac{\partial Y_i}{\partial w} = \left(\frac{\partial X_i}{\partial w} + \sum \frac{\partial E_{ij}}{\partial w_j} \right) + \sum \left(\frac{\partial Y_j}{\partial w} \cdot E_{ij} + \frac{\partial}{\partial F_{ij}^r} \left(\frac{\partial Y_j}{\partial w} \right) + \frac{\partial G_{ij}}{\partial w} \cap \frac{\partial Y_j}{\partial w} \right),$$

or simply $Y_i(w) = X_i(w) + \sum M_{ij}(w)[Y_j(w)]$, where $Y_i(w) = \partial Y_i/\partial w$, $X_i(w) = \partial X_i/\partial w + \sum \partial E_{ij}/\partial w_j$, $w_j = \partial w/\partial Y_j$, and

$$M_{ij}(w) = E_{ij}^r + {}^0 F_{ij}^r + ({}^0 G_{ij}/\partial w).$$

A little consideration shows that the events $Y_i(w)$ are in fact the minimal solutions of these equations, and so we have the matrix identity $Y(w) = (M(w))^*[X(w)]$. But the regularity of X_i, E_{ij}, G_{ij} implies that $M(w)$ and $X(w)$ take only finitely many values as w varies, so that the same must be true of $Y(w)$, and so the Y_i are indeed regular, having only finitely many distinct derivatives $Y_i(w) = \partial Y_i/\partial w$.

CHAPTER TEN

Context-free languages

Context-free languages are usually defined as follows, but in a slightly different terminology. We have two finite alphabets, of terminal and transient (or non-terminal) letters, and a grammar, which is a finite set of *productions* $A \rightarrow w$, in each of which A is a transient letter and w a word (which may involve letters of either type). A word involving at least one transient letter is called transient, while one involving only terminal letters is called terminal.

If u and v are words, and $A \rightarrow w$ a production of Γ, we write $uAv \rightarrow uwv$ (by Γ), and if as a consequence we have a chain $w_1 \rightarrow \ldots \rightarrow w_n$ (by Γ) we write $w_1 \rightarrow w_n$ (by Γ^*). (The value $n = 1$ is permitted.) The *terminal Γ-image* $\Gamma_{im}[w_0]$ of a word w_0 is defined as the set of all terminal words w for which $w_0 \rightarrow w$ (by Γ^*). An event E is called a *context-free language* if it is the terminal Γ-image of a transient letter under some grammar Γ. In this chapter we use the unqualified word *language*, and write **L** for the class of all languages.

We can easily put the definition into operator terminology. We let the terminal letters be $a, b, c, \ldots$ and the transient letters $A, B, C, \ldots$ and consider operators over the alphabet $V = \{a, b, c, \ldots, A, B, C, \ldots\}$. Then the production $A \rightarrow w$ has the same effect as the operator (w, A), and so Γ can be identified with the operator $\sum (w, A)$, summed over all productions of Γ. We then have $u \rightarrow v$ (by Γ) if and only if $v \in [u]$, and $u \rightarrow v$ (by Γ^*) if and only if $v \in \Gamma^*[u]$. So an event in $a, b, c, \ldots$ is a language if and only if it can be written in the form $(a + b + c + \ldots)^* \cap \Gamma^*[A_1]$, where Γ is of the form $(w_1, A_1) + \ldots$

$+ (w_n, A_n)$ and is defined over an extended alphabet $\{a, b, c, \ldots, A, B, C, \ldots\}$, in which the A_i are new letters. We shall return to the operator description later.

In the standard treatment the transient letters are construction letters used as scaffolding in forming the language, but then discarded. We propose to develop the theory in a less orthodox way, in which this scaffolding never appears. We directly characterize the terminal images of the transient letters in terms of certain equations they satisfy. The equations are obtained as follows. For any word w let $\bar{w}$ be obtained from w by replacing every transient letter A by a corresponding variable $\bar{A}$. Then if $A \rightarrow w_1, \ldots, A \rightarrow w_n$ are the productions from A, the equation corresponding to A is $\bar{A} = \bar{w}_1 + \ldots + \bar{w}_n$. Corresponding to Γ we get in this way a system $\bar{\Gamma}$ of equations.

As an example we consider the language $L = \Gamma_{im}[A]$, where Γ is the grammar

$$\{A \rightarrow BaC, \; B \rightarrow b, \; B \rightarrow AB, \; C \rightarrow a, \; C \rightarrow A\}.$$

Then working over the alphabet $\{a, b, A, B, C\}$ we can write $L = (a + b)^* \cap \Gamma^*[A]$, where Γ now denotes the operator

$$(BaC, \, A) + (b + AB, \, B) + (a + A, \, C).$$

The corresponding system of equations $\bar{\Gamma}$ is

$$\bar{A} = \bar{B}a\bar{C}, \qquad \bar{B} = b + \bar{A}\bar{B}, \qquad \bar{C} = a + \bar{A}.$$

We can see that these equations are satisfied by the terminal images of A, B, C as follows. Considering for instance the first time a production is applied to the letter B, we conclude that a word is in the terminal image of B if and only if it is *either b or* in the terminal image of AB. But since Γ^* is a homomorphism, the terminal image of AB is the product of those of A and B, and we deduce that the second equation of $\bar{\Gamma}$ holds for these terminal images. There might possibly be other solutions of the equations $\bar{\Gamma}$, but we can see from the argument that *any* solution satisfies

$$\bar{A} \geqslant \Gamma_{im}[A], \qquad \bar{B} \geqslant \Gamma_{im}[B], \qquad \bar{C} \geqslant \Gamma_{im}[C]$$

so that the system of terminal images is the *minimal solution* of $\bar{\Gamma}$.

In many simple cases we can 'solve' such equations one-by-one, expressing the languages they define as regular events or in terms of better known languages. We consider the minimal solution only, and use the fact that the minimal solution of $X = \alpha + \beta X$ is $X = \beta^* \alpha$. Solving the second equation of our example in this way, and then substituting from the second and third equations into the first, we get the equation $L = L^* b . a . (a + L) = L^* ba^2 + (L^* ba) . L$, which we can solve to $L = (L^* ba)^* . L^* ba^2 = (L + ba)^* . ba^2$. From this it follows that L is a function of ba and ba^2 only; and so, since $L + ba \geqslant ba^2 + ba$, we have $(L + ba)^* = (ba^2 + ba)^*$, and so $L = (ba^2 + ba)^* . ba^2$, a regular event.

Manipulations of this kind would be more difficult if performed directly in terms of grammars. They are justified by

Theorem 1. *To any system of inequalities*

$$X_1 \geqslant f_1(X_1, \ldots, X_n, E_1, \ldots, E_m)$$

$$\cdots$$

$$X_n \geqslant f_n(X_1, \ldots, X_n, E_1, \ldots, E_m)$$

*in which $f_1, \ldots, f_n$ are **S**-functions and $E_1, \ldots, E_m$ arbitrary events, there corresponds a second system*

$$X_1 \geqslant g_1(E_1, \ldots, E_m)$$

$$\cdots$$

$$X_n \geqslant g_n(E_1, \ldots, E_m).$$

*in which the g_i are **S**-functions (independent of the E_i). The second system is implied by the first, and is the minimal solution of the first, in the sense that if events $X_1, \ldots, X_n$ satisfy the second system with equality then they satisfy the first with equality also.*

Proof. We define sequences of approximating solutions X_{iN} and g_{iN} by the equations

$$X_{i0} = g_{i0}(E_1, \ldots, E_m) = 0$$

$$X_{i(N+1)} = g_{i(N+1)}(E_1, \ldots, E_m) = f_i(X_{1N}, \ldots, X_{nN}, E_1, \ldots, E_m).$$

Then we define $g_i(E_1, \ldots, E_m) = \sum g_{iN}(E_1, \ldots, E_m)$ over all N.

Obviously, if events X_i satisfy the first system we can prove inductively that $X_i \geqslant g_{iN}(E_1, \ldots, E_m)$ for all N, and so $X_i \geqslant g_i(E_1, \ldots, E_m)$. If, however, we *define* X_i as $g_i(E_1, \ldots, E_m)$, then we have

$$X_i = \sum X_{i(N+1)} = \sum f_i(X_{1N}, \ldots, X_{nN}, E_1, \ldots, E_m)$$

$$= f_i(X_i \ldots, X_n, E_1, \ldots, E_m)$$

since every word of $f_i(X_1,...,X_n,E_1,...,E_m)$ is in some summand $f_i(X_{1N},...,X_{nN},E_1,...,E_m)$. This shows that the X_i as defined satisfy the first system.

The theorem allows us to solve such systems piecemeal – we can solve some of the equations as a separate system, treating the other variables as constants, and substitute the partial solutions into the remaining equations. We can even partially solve an equation for a variable X, treating some of the occurrences of X as constants. In future, when performing such manipulations we shall leave it to the reader to verify that minimality is preserved at each stage.

The treatment in terms of equations is unusual, and we feel obliged to add some words of justification. We contend that the treatment is *logically simpler*, using only one concept where the usual approach has two – both transient latters and their terminal images. It is often *technically simpler* to manipulate equations rather than grammars, and the manipulations suggest themselves more readily. And in any case the translation from equations to grammars is so easy that a reader who prefers grammars can always translate the whole argument for himself.

Theorem 2. *We have* $L(L) = L$. *In other words, an* L*-function of languages is again a language. Also,* $R \subseteq L$, *so that* L *is closed under regular operations and contains all regular events.*

Proof. Let l be an L-function of n variables, and $L_1, \ldots, L_n$ be languages. Then $l(L_1,...,L_n)$ is the language defined by the system $\sum_0 \cup \sum_1 \cup ... \cup \sum_n$ of equations, where $\sum_1, \ldots, \sum_n$ are systems defining $L_1, \ldots, L_n$, and $\sum_0$ is obtainable from a system defining $l(x_1,...,x_n)$ on replacing each variable x_i by the corresponding L_i. The systems $\sum_0, \ldots, \sum_n$ are understood to have disjoint sets of variables. The second part of the theorem follows from the fact that the equations $W = 1$, $X = A + B$, $Y = AB$, $Z = 1 + AZ$ define the particular events $1, A + B, AB, A*$.

For example, we get a system of equations defining $L*$ on supplementing a system for L by a new equation $Z = 1 + LZ$. In view of the theorem, if all the events and functions appearing in a system of equations belong to the class L, then so do their minimal solutions.

Theorem 3. *The intersection of a language with a regular event is again a language.*

Proof. Let languages $X_1, \ldots, X_n$ be defined by the equations $X_s = f_s(\ldots, Y_l, \ldots)$, where the f_s are regular functions and the events Y_l consist of the events X_l together with the inputs $a, b, c, \ldots$. Let $E = E!(E_1, \ldots, E_k)$ be any regular event. Then we have

$$X_x \cap E_t = (f_s \wedge E_t!)(\ldots, Y_i \cap E_j, \ldots)$$

for by factor theory it is obvious that every word on either side of one of these equations appears on the other. The mn equations constitute a defining system for the events $X_s \cap E_t$.

In view of Theorems 2 and 3 and Theorem 6 of Chapter 8, we have

$$[R, L]_R . [R, L]_R = [R, L]_R \subseteq \langle R, R \rangle$$

$$[L, R]_R . [L, R]_R = [L, R]_R$$

$$[L, R]_R \subseteq \langle L, L \rangle$$

and

$$[L, S]_L \subseteq \langle L, R \rangle$$

with the new meaning for **L**. In particular, the class **L** is closed under all biregulators. So for instance the derivative of a language by a regular event, or its shuffle or bishuffle with a regular event, is again a language. Another powerful theorem relating the class **L** with operator theory is the one we promised in Chapter 9:

Theorem 4. *The open convex members of the class* $[L, R]_R$ *form a regular algebra.* (*The dual class is, of course, a regulator algebra.*)

Proof. Let Ω belong to the class concerned, so that it suffices to prove $\Omega^* \in [L, R]_R$. Now it is easy to see that Ω has the same star as $\Psi = (\Omega[1], 1) + (\Omega[a], a) + \ldots + (\Omega[k], k)$, where $\{a, b, \ldots, k\}$ is the alphabet, and that the events $X = \Psi^*[1], A = \Psi^*[a], \ldots, K = \Psi^*[k]$ are defined by the equations

$$X = (\chi(A, B, \ldots, K))^*$$

$$A = XaX + X.\alpha(A, B, \ldots, K).X$$

$$\cdots$$

$$K = XkX + X.\kappa(A, B, \ldots, K).X$$

where the L-functions χ, α, β, . . ., are those for which

$$\Omega[1] = \chi(a, b, \ldots, k), \quad \Omega[a] = \alpha(a, b, \ldots, k), \ldots, \Omega[k] = \kappa(a, b, \ldots, k).$$

So X, A, B, . . ., K belong to L, and Ψ^*, being a homomorphism, belongs to $[\mathbf{L}, \mathbf{R}]_\mathbf{R}$.

Grammatically linear languages

A system of equations in variables X_1, X_2, . . . is called *linear* if each right-hand side is a sum of terms each of the form A or $BX_i C$ wherein A, B, and C are independent of the X_i. It is *left-linear* if each C is 1, *right-linear* if each B is 1.

Theorem 5. *The minimal solutions of a system of left- or right-linear equations are regular functions of the parameters A, B, C. The minimal solutions of arbitrary linear equations can be written as sums of terms of the form $\Omega[A]$, wherein Ω is a regular function of the operators B^l, C^r.*

Proof. A family of left-linear equations can be written in matrix form $X_i = A_i + \sum B_{ij} X_j$, whose solution is $X_i = \sum D_{ij} A_j$, where the matrix (D_{ij}) is the star of (B_{ij}), and a similar proof applies to right-linear equations. But we can also write two-sided linear equations in matrix form by a technical device – we observe that

$$X_i = A_i + \sum \Omega_{ij}[X_j],$$

where each Ω_{ij} is a sum of operators of the form $B^l C^r$, and the theorem follows on taking the star of the operator matrix (Ω_{ij}).

The last part contains *Gruska's theorem*, that the languages defined by linear grammars (those in which every production has the form $A \to u$ or $A \to vBw$, with u, v, w terminal words) are precisely those which can be written in the form $f(a^l, a^r, b^l, b^r, \ldots)[1]$, where f is a regular function. This class – which is a proper subset of the context-free-languages – we call the *grammatically linear languages*. We showed in Chapter 9 that the operators $\mathbf{R}(\mathbf{R}^l, \mathbf{R}^r)$ were not all regulators – we now know that they belong to the class $\langle \mathbf{L}, \mathbf{L} \rangle$, and indeed take every grammatically linear language to another.

We should add that Gruska characterizes also the *weakly-bounded* languages, that is those generated by grammars with the property that if $A \to w$ (by Γ^*) then w involves at most a bounded number of occurrences of A. He considers (in a very different notation) the least classes **J**, **K** of languages and operators such that

(i) **J** and **K** are closed under regular functions
(ii) if $E \in$ **J** and $\Omega \in$ **K**, then $\Omega[E] \in$ **J** and E^l, $E^r \in$ **K**,

and shows that **J** is the class of weakly bounded languages.

The regular algebra generated by the operators a^l, a^r, b^l, b^r, ... we call *Gruska's algebra*, or the *algebra of multipliers*, and the algebra generated by A^l, A^r, B^l, B^r, ... is *the Gruska algebra generated by* A, B, ... Gruska actually considered an abstract algebra of sets of ordered pairs (his pair (u,v) corresponding to our $u^l v^r$), and there is another way in which this abstract algebra is embedded in our system (his pair (u,v) now corresponding to our $[u, \leftarrow v]$, where $\leftarrow v$ is the reverse of v). So the algebra of biregulators, with the biregular operations, is abstractly isomorphic to Gruska's algebra.

We set up the isomorphism as follows. Define

$$\|\Omega\| = \sum u^l v^r \text{ (over pairs } u, v \text{ with } [u, \leftarrow v] \in \Omega)$$

$$\Omega[\![X]\!] = \sum uXv \text{ (over the same set of pairs).}$$

Then we have

$$\|\Omega + \Psi\| = \|\Omega\| + \|\Psi\|, \qquad \|\Omega : \Psi\| = \|\Omega\| . \|\Psi\|$$

$$\|\Omega^{**}\| = \|\Omega\|^*, \qquad \Omega[\![X]\!] = \|\Omega\|[X].$$

Thus $\|\ \|$ converts the biregular operations to the regular ones, and the *biapplicator* $[\![\ ]\!]$ to the *applicator* $[\ ]$. The event $\Omega[\![X]\!]$ may be called the *biimage* of X under Ω. We can therefore express Gruska's theorem in the equivalent form that the grammatically linear languages are the biimages of 1 under biregulators, and we have proved that the biimage of any grammatically linear language under a biregulator is again grammatically linear.

We have already shown that the derivative of a language by a regular event is a language. In fact rather more is true. If $X_i = f_i(X_1, \ldots, X_k, a, b, c, \ldots)$ is a system of *regular equations* (i.e., the f_i are regular functions), we shall call the events $X_1, \ldots, X_k$ and the inputs a, b, c, ... the *companions* of X_1. (We do not mean to imply that X_1 has a unique set of companions.)

Theorem 6. *The word derivates of a language are regular functions of its companions.*

Proof. We can differentiate the equation $X_1 = f_1(X_1,...,X_k)$ (in which we have suppressed dependences on inputs) by the letter a to obtain an equation of the form

$$\frac{\partial X_1}{\partial a} = g(X_1,...,X_k) + \frac{\partial X_1}{\partial a}.h_1(X_1,...,X_k) + ... + \frac{\partial X_k}{\partial a}.h_k(X_1,...,X_k)$$

with regular functions g, h_1, . . ., h_k. In this way we obtain a system of right-linear equations for the derivates $\partial X_1/\partial a$, . . ., $\partial X_k/\partial a$, of which they are in fact the minimal solutions, and so can be expressed as regular functions of the X_i and the inputs. We obtain word derivates on repeatedly differentiating by letters in this way.

As an example, we take the language X defined by the equation $X = (aX + Xb)^* c$. We observe that $1 \notin X$, and so we can differentiate to get

$$\frac{\partial X}{\partial a} = X.Y^* c + \frac{\partial X}{\partial a}.bY^* c, \quad \frac{\partial X}{\partial b} = \frac{\partial X}{\partial b}.bY^* c, \quad \frac{\partial X}{\partial c} = 1 + \frac{\partial X}{\partial c}.bY^* c$$

where $Y = aX + Xb$. Solving these, and remembering that $Y^* c = X$, we get

$$\frac{\partial X}{\partial a} = X^2.(bX)^*, \quad \frac{\partial X}{\partial b} = 0, \quad \frac{\partial X}{\partial c} = (bX)^*.$$

We now have an effective test of whether a word w belongs to a given language L, since it is easy to decide whether $1 \in \partial L/\partial w$. The fact that derivates of languages by regular events are languages follows from our next theorem, which is the basis of factor theory for languages. Recall that the operator E^{lr} is defined by $E^{lr}[X] = \sum uXv$ over all pairs u, v with $uv \in E$.

Theorem 7. *If X is a language with companions $X, Y, Z, . . .$, then the operators $X^{lr}, Y^{lr}, . . .$, are regular functions of $X, X^r, Y, Y^r, ...$.*

Proof. Let the equation defining X be $X = f(X,Y,Z,...)$ with f regular, and let $x, y, z, . . .$ be the distinct input letters. Then by factor theory we can write $f(x,y,z,\ldots) = \sum l_i.r_i + \sum \lambda_i.x_i.\rho_i$, where $l_i, r_i, \lambda_i, \rho_i$ are regular functions of $x, y, z, \ldots$ with the properties

(i) if $uv \in f(x,y,\ldots)$ then $u \in \rho_i$, $v \in r_i$ or $u \in \lambda_i$, $v \in x_i\rho_i$ or $u \in \lambda_i x_i$, $v \in \rho_i$ for some i.

(ii) if $utv \in f(x,y,\ldots)$ and t is one of $x, y, z, \ldots$ then we have $u \in \lambda_i$, $t = x_i$, $v \in \rho_i$ for some i.

It then follows that we have

$$X^{lr} = \sum L_i^l.R_i^r + \sum \Lambda_i^l.P_i^r.X_i^{lr}$$

where $L_i, R_i, \Lambda_i, P_i, X_i$ are obtained from $l_i, r_i, \lambda_i, \rho_i, x_i$ on replacing each of $x, y, z, \ldots$ by the corresponding one of $X, Y, Z, \ldots$. We can write this equation in the form

$$X^{lr} = \Omega + \Omega_1 X^{lr} + \Omega_2 Y^{lr} + \ldots$$

in which the Ωs are functions of the $L_i^l, R_i^r, \Lambda_i^l, P_i^r$, when we see that it is left-linear in $X^{lr}, Y^{lr}, \ldots$. So it and the similar equations for $Y^{lr}, Z^{lr}, \ldots$ define these operators as regular functions of $X^l, X^r, Y^l, X^r, \ldots$ as required.

Using the isomorphism $\| \ \|$ we can alternately write this result as follows. Recall that $L^{l \leftarrow r}$ is the operator $\sum [u, \leftarrow v]$ $(uv \in L)$. Then if L is a language L^{l-r} is a biregular function of the operators $[M, 1]$ and $[1, \leftarrow M]$ as M ranges over the companions of L, and so belongs to the operator class $[\mathbf{L}, \mathbf{L}]_{\mathbf{R}}$. Since $L^{l \leftarrow r} = {}_\partial L^r . \leftarrow$, we can say that ${}_\partial L$ and ${}_\partial L^r$ belong to $[\mathbf{L}, \mathbf{L}]_{\mathbf{R}}$.

The theorems 'Ω biregular implies $\Omega[\![1]\!]$ a language', and 'L a language implies $L^{l \leftarrow r} \in [\mathbf{L}, \mathbf{L}]_{\mathbf{R}}$', are in a sense dual. We can often use them to prove that certain operators are not biregulators, or certain events not languages, by interchanging the two problems. We give some examples:

(i) The event $L = \sum a^n b^n a^n b^n$ (over $n \geqslant 0$) is not a language, since otherwise $(a^*b^*)^\cap . L^{l \leftarrow r} . (a^*b^*)^\cap \in [\mathbf{L}, \mathbf{L}]_{\mathbf{R}}$, contradicting the result of Chapter 8, since this operator is $\sum [a^n b^n, b^n a^n]$.

(ii) The operator $\frac{1}{2}$, defined by linearity and $\frac{1}{2}[uv] = u$ if u and v have the same length, is a regulator. To see this, observe that $\frac{1}{2}[E] = \sum L_i \cap [I,I]^{**} [R_i]$ if $L_i . R_i$ is the typical 2-term factorization of E and I the alphabet. But $\frac{1}{2}$ is not a biregulator, since if L is the language $\sum a^n b^n c^m d^2 e^{3m}$ (over m, $n \geqslant 0$) then $\frac{1}{2}[L] \cap a^* b^* c^* d = \sum a^n b^n c^n d$, so $\frac{1}{2}[L]$ is not a language, since we can easily prove that $\sum a^n b^n c^n$ is not a language.

(iii) Alternative proofs that $\bigcirc$ and $\leftarrow$ are not biregulators follow from the facts that $\bigcirc[\![1]\!]$ and $\leftarrow[\![1]\!] = \sum w^2$ are not languages, as is well known.

Radical algebras

It would seem that a mathematically natural setting for the theory of context-free languages is the following. We define a (left) *radical algebra* as a regular algebra of events which is closed under (left) differentiation by words and contains the letters of the alphabet. *Mnemonic*: *R*egular *A*lphabetic *DI*fferentially *C*losed *AL*gebra: also a radical algebra is one defined by the *roots* of equations, as we shall see. A *basis* for an algebra is a finite set of elements generating it as a regular algebra.

Theorem 8. *A left radical algebra with a finite basis is also a radical algebra. Accordingly, we use the unqualified term* radical algebra *for a left or right radical algebra with finite basis. The companions of any language generate a radical algebra, and every radical algebra is of this form. Every radical algebra is closed under left or right differentiation by regular events. If L belongs to a radical algebra then L^{lr} is a regular function of operators M^l, M^r (M in the algebra), and $L^{l \leftarrow r}$ is a biregular function of operators $[M, 1]$ and $[1, \leftarrow M]$, M in the algebra.*

Proof. Let $X, Y, Z, \ldots$ be a finite basis for a left-radical algebra. Then $\partial X / \partial a$, $\partial X / \partial b$, $\ldots$ are regular functions of $X, Y, Z, \ldots$ and so the equation

$$X = (0 \text{ or } 1) + a \cdot \frac{\partial X}{\partial a} + b \cdot \frac{\partial X}{\partial b} + \ldots$$

and the similar equations defining $Y, Z, \ldots$ prove that $X, Y, Z, \ldots$ are a system of companions for X, which is therefore a language. The right letter derivates of $X, Y, Z, \ldots$ are therefore also in the algebra, which is accordingly a right-radical algebra. The remaining parts of the theorem are translations of earlier results of this chapter.

The u, v, w, x, y lemma

Theorem 9. *If L is an infinite language, there are words u, v, w, x, y with $v \neq 1$, $x \neq 1$ such that $uv^n wx^n y \in L$ for all $n \geqslant 0$.*

Proof. We express L^{lr} as a regular function of operators M^l, M^r, M the typical companion of L, and consider the outermost starred parts of this expression, so that L^{lr} appears as a sum of terms of the form $\Omega_0 \Psi_1^* \Omega_1 \Psi_2^* \ldots \Omega_n$, wherein each Ω_i, being a finite product of factors M^l or M^r, reduces to the form $X^l Y^r$, and we can suppose no Ω_i is zero.

Now if an operator Ψ contains $v_0^l y_0^r$ with $v_0 \neq 1$ and a term $v_1^l y_1^r$ with $y_1 \neq 1$, then Ω^* contains their product, say $v^l y^r$ with $v \neq 1$, $y \neq 1$, and so contains $(v^n)^l (y^n)^r$ for all $n \geqslant 0$. It follows that a product $\Omega \Psi^* \chi$ with Ω and $\chi \neq 0$ contains $(uv^n w)^l (xy^n z)^r$ for all n, u, w, x, z being suitably chosen words. It follows that the lemma holds of L unless every Ψ_i in the expression for L^r reduces to the form X^l or X^r, when Ψ^* reduces to $(X^*)^l$ or $(X^*)^r$, and the expression for L^{lr} becomes a finite sum of terms $Y^l Z^r$, proving that L has finitely many derivates and so is regular. The lemma holds of infinite regular events, since if $uv^n w \in L$ for all n so is $uv^n vv^n w$.

The lemma emerges in this light as a shadow of the factor theorem for languages, Theorem 7, which gives more precise information of the same kind. It is theoretically important in that it gives an algorithm for deciding whether or not a language is infinite, since the words u, v, w, x, y can be effectively found from a grammar if L is infinite, and they visibly demonstrate its infinitude. The lemma also gives a quick proof that certain events – for instance $\sum a^n b^n c^n$ – are not languages, for the required u, v, w, x, y are lacking.

Various other simple problems for languages are known to be insoluble, for instance the problem of deciding whether two languages intersect, or whether a language is a regular event. It is also impossible to produce an algorithm which, given that a language is regular, produces a regular expression for it, although if we are given a system of regular equations $X_i = f_i(X_1, \ldots, X_n)$ and told that all the events X_i of the minimal solution are regular, then we can find regular expressions for them.

The equation $\sum a^n b^n . a^* \cap a^* . \sum b^n a^n = \sum a^n b^n a^n$ shows that the intersection of two languages need not be a language. There is an elegant example, due to Ginsburg and Spanier, showing that $\partial L/\partial M$ need not be a language for languages L, M. We let L, M be defined by

$$L = aLb^2 + bLa^3 + cLcba + d$$
$$M = aMa + bMb + cMc + d$$

Then $\partial L/\partial M$ is the event

$$ba + a^4 + a^3 b^2 + a^2 b^4 + ab^6 + b^8 + b^7 a^3 + b^6 a^6 + \ldots + a^{24} + \ldots$$

and $\partial L/\partial M \cap a^* = \sum a^{4 \cdot 6^n}$, proving that $\partial L/\partial M$ is not a language. (This follows from Parikh's Theorem – see Chapter 11.) It follows that operators of class $[\mathbf{L}, \mathbf{L}]_\mathbf{R}$ do not preserve languages, and that the Pilling product theorem cannot be generalized to prove

$$[\mathbf{L}, \mathbf{L}]_\mathbf{R} . [\mathbf{L}, \mathbf{L}]_\mathbf{R} \subseteq [\mathbf{L}, \mathbf{L}]_\mathbf{R}.$$

CHAPTER ELEVEN

Commutative regular algebra

If we repeat the definition of regular event, but regard the input letters as mutually commuting, so that they generate a free commutative semigroup (with unit), we obtain the definition of *commutative regular event*. Every *word* in the input letter a, b, c, . . . can now be reduced to the form $a^\alpha b^\beta c^\gamma$ . . . ($\alpha, \beta, \gamma, \ldots \geqslant 0$), and an *event* is an arbitrary set of such words. It is *regular* if it can be obtained from 0, 1, and the inputs by a finite number of applications of the regular operations $+$, $.$, $*$ (defined as usual).

Redko has given a system of axioms for commutative regular events. Owing to slightly different conventions, he omits some of the axioms concerning 0 and 1, and when these are added his system is essentially the system of *classical axioms* $C1$–14 (Chapter 3), supplemented by axioms C^+1, C^+2, C^+3 below. We shall see that C^+3 is redundant, and so we take $C1$–14, C^+1, C^+2 as our system of axioms for commutative regular algebra. We make deductions from these as in Chapter 4, using the results of that chapter freely.

C^+1 $AB = BA$

C^+2 $A^* B^* = (AB)^* (A^* + B^*)$

C^+3 $(A + B)^* = A^* B^*$

(To prove C^+3, put A^*, B^* for A, B in C^+2 and use $C13$ ($X^{**} = X^*$) to give $A^* B^* = (A^* B^*)^* (A^* + B^*) \geqslant (A + B)^* \geqslant A^* B^*$.)

C^+4 $(A^* B)^* = 1 + A^* B^* B$

(Since $(A^* B)^* = 1 + (A^* B)^* A^* B = 1 + (A + B)^* B = 1 + A^* B^* B$, using $C11$.)

91

Normal form

Theorem 1. *Any regular expression in $a, b, c, \ldots$ can be expressed in* normal form *as a finite sum of terms of the shape $w_1^* w_2^* \ldots w_n^* w$, where $w_1, \ldots, w_n, w$ are words (i.e., products) of the letters $a, b, c, \ldots$.*

Proof. We can use C^+3 to eliminate additions within stars, and then C^+4 (and the commutative law C^+1) to eliminate stars within stars. After this, every starred item is a word, and we use the commutative law to collect starred items at the end.

Now let $s = (\alpha, \beta, \ldots)$ be a finite sequence of numbers $\alpha, \beta, \ldots \geqslant 0$, not all zero.

$$C^+5.\text{s} \quad A^* B^* C^* \ldots = (A^\alpha B^\beta \ldots)^*$$
$$\times (A^{<\alpha} B^* C^* \ldots + A^* B^{<\alpha} C^* \ldots + \ldots).$$

We prove the 2-variable case $A^* B^* = (A^\alpha B^\beta)^* (A^{<\alpha} B^* + A^* B^{<\beta})$ by replacing A^*, B^* by $A^{\alpha^*} A^{<\alpha}$, $B^{\beta^*} B^{<\beta}$ and using C^+2 on $A^{\alpha^*} B^{\beta^*}$. This proof is valid only if $\alpha\beta > 0$, but if just one of α, β is zero the proof is simpler. The general case follows similarly from

$$A^* B^* C^* \ldots = (ABC \ldots)^* (B^* C^* \ldots + A^* C^* \ldots + \ldots),$$

which is proved by repeated use of C^+2.

Independent form

We call $w_1, \ldots, w_n$ *independent* if the equation

$$w_1^{\alpha_1} \ldots w_n^{\alpha_n} = w_1^{\beta_1} \ldots w_n^{\beta_n}$$

in non-negative integers α_i, β_i, implies $\alpha_1 = \beta_1, \ldots, \alpha_n = \beta_n$.

Theorem 2. *Every regular expression in $a, b, c, \ldots$ can be expressed in* independent form *that is, in a normal form for which the words $w_1, \ldots, w_n$ in each term $w_1^* \ldots w_n^* w$ are independent.*

Proof. If $w_1, \ldots, w_n$ are independent we can renumber them so that there is an equation $w_1^\alpha w_2^\beta \ldots = w_r^\gamma w_{r+1}^\delta \ldots$ for some non-negative

$\alpha, \beta, \ldots, \gamma, \delta, \ldots$, not all zero. But if we have $A^\alpha B^\beta \ldots = C^\gamma D^\delta \ldots = X$, say, then by C^+5 we have

$$A^* B^* \ldots = X^*(A^{<\alpha} B^* \ldots + A^* B^{<\beta} \ldots + \ldots)$$

and

$$C^* D^* \ldots = X^*(C^{<\gamma} D^* \ldots + C^* D^{<\delta} \ldots + \ldots)$$

and if we multiply these two expressions and use $X^* X^* = X^*$ we obtain an expression for $A^* B^* \ldots C^* D^* \ldots$ as a sum of terms each of which involves one less star than the original. We treat the product $w_1^* w_2^* \ldots w_r^* w_{r+1}^* \ldots w$ in this way, and continue until all dependencies have been removed.

Now our aim is to prove *Redko's theorem*, that every identity between regular expressions which is valid for all commutative events (i.e., is a *commutative tautology*) is a consequence of the axioms. We write $A \subseteq B$ to mean that the event represented by A is contained in that represented by B, and $A \cap B = 0$ to mean that these two events are disjoint. We can then express the theorem in the form '$A \subseteq B$ implies $A \leqslant B$' (the latter meaning that $A + B = B$ follows from the axioms), since then if A and B represent the same event we shall have $A \leqslant B \leqslant A$, and so $A = B$, as a consequence of the axioms.

We suppose we have an independent term $T = t_1^* t_2^* \ldots t_n^* t$, which will be kept fixed throughout the discussion.

Theorem 3. *Any regular expression X can be expressed via the axioms as $X_1 + X_2$, in which $X_1 \leqslant T$, $X_2 \cap T = 0$.*

Proof. The proof is a long one, and involves several lemmas. We observe first that if X is a sum $Y + Z + \ldots$ of terms for which the theorem holds, then we can write $X_1 = Y_1 + Z_1 + \ldots$, $X_2 = Y_2 + Z_2 + \ldots$ for the corresponding decompositions, and deduce the theorem for X. So we shall suppose that X is a single normal term.

We now find it convenient to introduce *fractional words*, defined as formal products $a^\alpha b^\beta \ldots k$, where $\alpha, \beta, \ldots$, are arbitrary rational numbers. The *fractional power* w^α of a word (α being any rational number), and the *product* vw of two fractional words, are defined as one would expect. Then of course the set of all fractional words is a vector space of dimension N, the number of letters $a, b, \ldots, k$, but written multiplicatively instead of additively. The words $t_1, \ldots, t_n$ of our term T, being independent, are part of a base for this space, say $t_1, \ldots, t_n, t_{n+1}, \ldots, t_N$. Any word w has a unique expression $w = t_1^{\alpha_1} \ldots t_N^{\alpha_N}$ with rational α_i – we call $w t_i$-*positive* or t_i-*negative* or t_i-*zero* according as α_i is positive or negative or zero. A term $w_1^* \ldots w_m^* w$ is called t_i-*mixed* if there exists among $w_1, \ldots, w_m$ both a t_i-positive and a t_i-negative word.

Lemma 1. *Any term $U = u^* v^* \ldots w$ can be expressed as a finite sum of terms each of which is t_i-unmixed for every i. We shall call such a term* unmixed.

Proof. Suppose say that U is t_1-mixed, with u being t_1-positive and v t_1-negative. Then for suitable positive integers α and β the word $u^\alpha v^\beta$ will be t_1-zero. If we replace $u^* v^* \ldots w$ by

$$(u^\alpha v^\beta)^* (u^{<\alpha} v^* + u^* v^{<\beta}) \ldots w$$

we have replaced a term with starred words $u, v, \ldots$ by a sum of terms with starred words either $u^\alpha v^\beta, v, \ldots$ or $u, u^\alpha v^\beta, \ldots$, and so increased the number of t_1-zero starred words per term, without increasing the total number of starred words per term. Continuing, we eventually express U as a sum of t_1-unmixed terms. We then apply the same unmixing process with $t_2, \ldots, t_N$, observing that t_i-unmixed terms remain t_i-unmixed during the t_j-unmixing process ($j > i$), so that ultimately every term is t_i-unmixed for every i.

Lemma 2. *Any unmixed term U with a t_i-negative starred word ($i = 1, \ldots, N$), or a t_j-positive starred word ($j = n+1, \ldots, N$) can be expressed as a finite sum of unmixed terms U_i each of which either has fewer starred words or satisfies $U_i \cap T = 0$.*

Proof. Let $U = u^* v^* \ldots w$ be unmixed, with $u t_1$-negative (say). We expand $t = t_1^{\alpha_1} \ldots t_N^{\alpha_N}$, $u = t_1^{\beta_1} \ldots t_N^{\beta_N}$, $w = t_1^{\gamma_1} \ldots t_N^{\gamma_N}$ in terms of our base, and consider the corresponding expansions of the typical words of U and T. The index of t_1 in the expansion of $u^\alpha v^\beta \ldots w$ is at most $\alpha \beta_1 + \gamma_1$, while the index of t_1 in any term of T is at least α_1. Since β_1 is negative by assumption, for sufficiently large α we have $u^\alpha v^\beta \ldots w \notin T$, so that for some α the term $U_\alpha = u^\alpha u^* v^* \ldots w$ satisfies $U_\alpha \cap T = 0$. We now expand U as $u^\alpha u^* v^* \ldots w + u^{<\alpha} v^* \ldots w$, proving the lemma in this case. The other cases are similar.

After repeatedly applying the process of Lemma 2 to the decomposition given by Lemma 1 we have a sum of terms U_i each of which *either* satisfies $U_i \cap T = 0$ *or* contains only t_i-positive or t_i-zero words for $i = 1, \ldots, n$, and t_j-zero words for $f = n + 1, \ldots, N$. But if a word u satisfies these conditions, then for some positive integer α we have $u^\alpha = t_1^{\alpha_1} \ldots t_n^{\alpha_n}$, in which the α_i are non-negative integers. For any such u we replace a factor u^*, wherever it appears, by $u^{\alpha^*} u^{<\alpha}$. After this, we have a sum of terms $U_i = u^* v^* \ldots w$ each of which *either* satisfies $U_i T = 0$ *or* has a word $u^\alpha v^\beta \ldots w$ in common with T *and* is such that each of u, v, $\ldots$ is a product of the words $t_1, \ldots, t_n$ (with non-negative integral indices). In the latter kind of term we now replace $u^*, v^*, \ldots$ by $u^\alpha u^* + u^{<\alpha}, v^\beta v^* + v^{<\beta}, \ldots$, and get a sum of terms with fewer starred words and the term $u^* v^* \ldots u^\alpha v^\beta \ldots w = U'$, say. Since the words $u, v, \ldots$ are products of $t_1, \ldots, t_n$, while $u^\alpha v^\beta \ldots w \in T$, we can deduce from the axioms that $U' \leqslant T$. So Theorem 3 is proved (at last!) by induction on the number of starred words per term.

Theorem 4. *If X and Y are regular expressions with $X \subseteq Y$, then $X \leqslant Y$.*

Proof. We express $Y = T_1 + T_2 + \ldots + T_k$ in independent form. Using Theorem 3, we express X as $X_1 + X'$, with $X_1 \leqslant T_1$, $X' \cap T_1 = 0$. Then we express X' as $X_2 + X''$, with $X_2 \leqslant T_2$, and $X'' \cap T_2 = 0$. Continuing, we obtain an expression $X = X_1 + X_2 + \ldots + X_k + X^{(k)}$, in which $X_i \leqslant T_i$ and $X^{(k)} \cap T_i = 0$ for each i. But since $X \subseteq Y$ this implies that the event represented by $X^{(k)}$, being contained in X and disjoint from Y, must be empty, so that we have $X^{(k)} = 0$, and $X \leqslant Y$.

We have already pointed out that this contains Redko's theorem. As Pilling has pointed out, the published proofs of both Redko and Solomaa are incomplete. These authors both assert that the problem reduces to the case in which all starred items in one of the two expressions are single letters, and they give valid inductive proofs in this case. The content of these proofs is essentially the argument summarized in the penultimate sentence before our Theorem 4. But the reduction to the special case is not at all obvious, and it is this reduction which occupies the major part of our proof. The first complete proof would seem to be that given by Pilling. Commutative regular algebra is notable for the number of results whose proofs one would expect to be trivial, but which turn out to be very subtle.

The proof of Redko's theorem shows in passing that commutative regular events are closed under the Boolean operations, but there are simpler proofs of this. For *intersections*, it suffices to consider the intersection $V \cap W$ of two normal terms $V = v_1^* \ldots v_n^* v$ and $W = w_1^* \ldots w_m^* x$. If this is finite, it is obviously regular. If not, applying Higman's theorem to the set of abstract words $x_1^{\alpha_1} \ldots a_n^{\alpha_n} . y_1^{\beta_1} \ldots y_m^{\beta_m}$ in $m + n$ letters for which $v_1^{\alpha_1} \ldots v_n^{\alpha_n} . v = w_1^{\beta_1} \ldots w_m^{\beta_m} . w$ we conclude that there are two such equalities, say $v_1^{\alpha_1} \ldots v_n^{\alpha_n} . v = w_1^{\beta_1} \ldots w_m^{\beta_m} . w$ and $v_1^{\gamma_1} \ldots v_n^{\gamma_n} . v = w_1^{\delta_1} \ldots w_m^{\delta_m} . w$ satisfying $\alpha_i \leqslant \gamma_i . \beta_j \leqslant \delta_j$. From these, by division, we get that some non-trivial product of the v_i coincides with a product of the w_i.

If this product is $t = v_1^{\alpha_1} \ldots v_n^{\alpha_n}$ (in a new notation), then using C^+5 we can express V as a sum of terms in each one of which t is a starred word (and the total number of starred words per term is unchanged). Similarly, expressing W, we see that our problem reduces to the computation of $t^* A \cap t^* B$. But $t^N A \cap t^M B$ is $t^N(A \cap t^{M-N} B)$ or $t^M(t^{N-M} A \cap B)$ according as $M \geqslant N$ or $M \leqslant N$, and so we have the formula

$$t^* A \cap t^* B = t^*(A \cap t^* B + t^* A \cap B),$$

and in each of the intersections within the bracket the total number of starred words has been reduced, so that we have two simpler problems. For the process to be effective we need a test determining whether the intersection $v_1^* \ldots v_n^* \cap w_1^* \ldots w_n^*$ contains a non-trivial word, but this amounts only to the solution of some linear inequalities in integers.

For *Boolean differences* there is a similar method based on the formula

$$t^* A \setminus t^* B = t^*(A \setminus B)$$

which is valid for independent terms $t^* A$ and $t^* B$ with $A \geqslant B$. To find the general difference $X \setminus Y$ we first reduce to the case when X is a single independent term, then replace Y by $X \cap Y$, so that we can suppose $Y \leqslant X$. If $X \cap Y$ is finite the problem is trivial, and if not we can show as above that we can take $X = t^* A$, $Y = t^* B$ for some word t, without increasing the number of stars on either side. Then let $A = v_2^* \ldots v_n^* v$ and $B = w_2^* \ldots w_m^* w$. Then since $X \geqslant Y$ we have $w = t^{\alpha_1} . v_2^{\alpha_2} \ldots v_n^{\alpha_n} . v$ say, and on replacing $t^*, v_2^*, \ldots, v_n^*$ by $t^{<\alpha_1} + t^{\alpha_1} . t^*, \ldots, v_n^{<\alpha_n} + v_n^{\alpha_n} . v_n^*$ in X we get an expression for X as a sum of terms with fewer starred words (for which we regard the problem as solved) and the term $t*v_2^* \ldots v_n^* w$, so that our problem reduces to finding the difference $t*v_2^* \ldots v_n^* \backslash t*w_2^* \ldots w_m^*$. But since the left side contains the right each of the w_i is a product of terms chosen from t and the v_i, say $w_i = t^{\alpha_i} . u_i$, where $u_i \in v_2^* \ldots v_n^*$. Then

$$t*v_2^* \ldots v_n^* \backslash t*w_2^* \ldots w_m^* =$$
$$= t*v_2^* \ldots v_n^* \backslash t*u_2^* \ldots u_m^* + t*u_2^* \ldots u_m^* \backslash t*w_2^* \ldots w_m^*.$$

The first of these satisfies the conditions for the applicability of the formula $t*A \backslash t*B = t*(A \backslash B)$, and the second is simpler than the first (and so regarded as solved) if $m < n$. If $m = n$, we can replace $t*u_2^* \ldots u_m^*$ (using the rule $Z^* = Z^{<\alpha} + Z^{\alpha} Z^*$) by a sum of terms with fewer starred words and the term $t*w_2^* \ldots w_m^*$ itself, and so the difference $t*u_2^* \ldots u_m^* \backslash t*w_2^* \ldots w_m^*$ is regular in this case also. This ingenious argument is due to Pilling.

Biregulators in the commutative case

Biregulators and the biregular operations are defined just as in the non-commutative case. Since $[a,1]$ and $[1,b]$ commute under bi-products, as do $[a,1]$ and $[b,1]$, the algebra of biregulators in n letters under the biregular operations is abstractly isomorphic to the free commutative regular algebra on $2n$ letters. In particular, every biregulator can be expressed in *normal form* as a sum of terms

$$[u_1, v_1]^{**} : \ldots : [u_m, v_m]^{**} : [u, v]$$

in which $u_1, v_1, \ldots, u_m, v_m, u, v$ are words.

Theorem 5. *The product $\Omega\Psi$ of two biregulators is a third. Every biregulator is a regulator.*

Proof. The proof mimics our construction for the intersection of regular events. We consider the operators

$$\Omega = [u_1, v_1]^{**} : \ldots : [u_n, v_n]^{**} : [u, v]$$

and

$$\Psi = [w_1, t_1]^{**} : \ldots : [w_m, t_m]^{**} : [w, t].$$

If the intersection $v_1^* \ldots v_n^* v \cap w_1^* \ldots w_m^* w$ is finite, we have nothing to fear. If not, we can suppose as in the intersection method that $v_1 = w_1 = T$, say. We now use the formula

$$([u_1, T]^{**} : \Gamma) . ([T, t_1]^{**} : \Delta) = [u_1, t_1]^{**} : \Phi,$$

where $\Phi = ([u_1,T]^{**}:\Gamma).\Delta + \Gamma.([T,t_1]^{**}:\Delta)$, which is proved exactly as the intersection formula. The second part of the theorem follows from the first, on taking for Ω an operator of the form $[X,1]$, X regular.

The form of this proof differs from that of the non-commutative analogue. It is *not* true in the commutative case that the class of regulators is closed under the biregular operations. Let Ω be the operator $\sum [1,a^p]$ over prime p. Then Ω is a regulator, since its only values are 0 and 1. But $(\Omega:[b,b]^{**})[(ab)^*] = \sum b^p$, an irregular event. However, the commutative theory of context-free languages is fairly simple, and this implies that certain theorems on operators hold also in the commutative case.

The key theorem is *Parikh's theorem*, that the commutative image of a context-free language is regular. The editors of the Journal in which this first appeared describe the theorem in a footnote as 'among the most fundamental and subtly difficult to prove in the theory (of context-free languages)'. We agree with that view, and point out that the subtlety is of the same kind as that on which we have already remarked in Redko's theorem and the Boolean closure property. Even when very short proofs of such theorems exist, they can be astonishingly difficult to find. We give here Pilling's short proof of the Parikh theorem, which for us is most naturally stated in the form below.

Theorem 6. *The minimal solutions of a set of regular equations* $X_i = f_i(X_1,\ldots,X_n,A_1,\ldots,A_m)$ $(i=1,\ldots,n)$ *are regular.*

Proof. We consider a single equation $X = f(X,A,B,\ldots)$, or, suppressing the dependence on parameters, $X = f(X)$. Using the axioms, we can rewrite this in the form $X = E + F(X).X$, wherein E is independent of X. It follows that

$$X \geqslant (F(X))^* E \geqslant (F(E))^* E = G^* E, \text{ say.}$$

But $G^* E$ is in fact a solution, therefore the minimal one, of $X = E + F(X).X$. To see this, we remark that if $\varphi(X,A,B,\ldots)$ is any word in $X, A, B, \ldots$ which involves X, then

$$\varphi(G^* E, A, B,\ldots) = G^* \varphi(E, A, B,\ldots)$$

using $G^* G^* = G^*$ and the commutative law. So

$$E + F(G^* E).G^* E = E + G^*.F(E).E = E + G^*.G.E = G^* E$$

proving our assertion.

We solve systems with more than one variable using this as an induction step and eliminating the variables one by one. We copy Pilling's example:

$$X = X^3 Ya + XY^2 a + b$$

$$Y = XY^2 + a.$$

The Y-equation can be written $Y = [a] + [XY] Y$, and so has the solution $Y = (Xa)^* a$. Substituting this in the X-equation, we get $X = X^3(Xa)^* a^2 + X(Xa)^* a^3 + b$, or in the required form

$$X = [b] + [(Xa)^* X^2 a^2 + (Xa)^* a^3] X,$$

whose solution is

$$X = [(ba)^* b^2 a^2 + (ba)^* a^3]^* b,$$

whence from $Y = (Xa)^* a$ we get

$$Y = [[(ba)^* b^2 a^2 + (ba)^* a^3]^* ba]^* a.$$

Expressed in normal form, these solutions become

$$X = b + (ba)^* (a^3)^* a^2 b^3 + (ba)^* (a^3)^* a^3 b$$

$$Y = a + (ba)^* (a^3)^* a^2 b.$$

Theorem 7. *The open convex biregulators form a regulator algebra, in the commutative case.*

Proof. The proof is just like that of Theorem 4 of the last chapter, but in the commutative case we can now replace **L** by **R**.

Theorem 7 does not appear to be even nearly the whole truth. Pilling conjectures (*the open mapping conjecture*) that the class of *all* open biregulators is a regulator algebra, and provides a mass of evidence to support this view. The conjecture is equivalently phrased in the form 'the star of an open biregulator is another'. In view of the method of proof of Theorem 7, we can regard the conjecture as a generalization of Parikh's theorem, and the reader who was inclined to doubt our remarks on the subtlety of that and similar theorems might welcome the opportunity to prove his point.

CHAPTER TWELVE

Some axiomatic questions

For the reader's convenience, we recall some definitions from Chapter 4. **S**-algebras should by now be familiar concepts – they have operations $\sum$, ., * satisfying the obvious laws, with $x^* = \sum x^n$. **N**-algebras are subsets of **S**-algebras closed under $+$, ., *. **R**-algebras are formal systems with operators $+$, ., * satisfying all the laws which are satisfied in all **N**-algebras, and **C**- and **A**-algebras are similar systems in which $+$, ., * are required to satisfy only $C1$–14, $C1$–13 respectively.

Finite algebras

In a finite algebra of any of these types the elements form a lattice, with $A \leqslant B$ meaning $A + B = B$, for we can define $A \cap B$ as the sum of all C with $C \leqslant A$, $C \leqslant B$. The addition-table is determined by the lattice ($A + B$ is the least C with $C \geqslant A$, $C \geqslant B$), and the star-table is determined when we indicate which elements are stars (A^* is the least star $\geqslant A$). The multiplication-table is then restricted by the conditions

$$A \leqslant B \text{ implies } AC \leqslant BC, CA \leqslant CB$$

among others, and in small cases this leave only a few possibilities. We conclude that all **C**-algebras of order 4 or less are among those listed, the printed table being the multiplication table for elements other than 0 and 1.

It is not obvious how we can test whether such a structure is an **R**-algebra, but it is easy to see that we can test whether a finite algebra

100

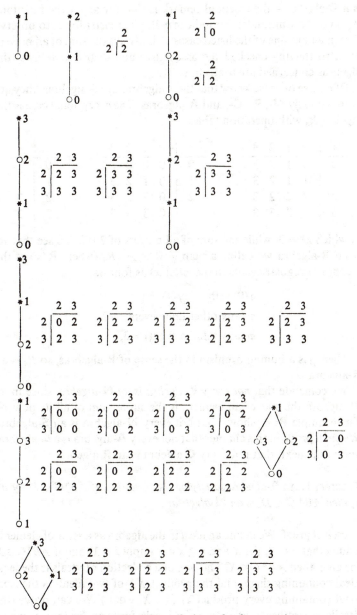

Fig. 12.1. The Kleene algebras of order up to 4.

is a **C**-algebra – the essential remark is that for any x the functions $x^{<n}$ and x^n are ultimately periodic in n. But it is easier for us to observe that in all but one of the listed cases x^* is always the sum of all powers of x, so (having checked the associative and distributive laws) the algebras concerned are **S**-algebras.

Of course once we know that these algebras are **S**-algebras, they are automatically **N**-, **R**-, **C**-, and **A**-algebras. The exceptional case is the algebra R_4 with operation tables

+	0	1	2	3
0	0	1	2	3
1	1	1	2	3
2	2	2	2	3
3	3	3	3	3

.	0	1	2	3
0	0	0	0	0
1	0	1	2	3
2	0	2	2	3
3	0	3	3	3

	*
0	1
1	1
2	3
3	3

in which $2^* = 3$, while the sum of all powers of 2 is 2. To see that R_4 is an **R**-algebra we define a map $\varphi : \mathbf{R}\langle x \rangle \to R_4$ (where $\mathbf{R}\langle x \rangle$ is the algebra of regular events in the letter x) as follows:

$$\varphi(0) = 0, \qquad \varphi(1) = 1$$
$$\varphi \text{ (any other finite event)} = 2$$
$$\varphi \text{ (any infinite event)} = 3,$$

Then φ is a homomorphism in the sense of **R**-algebras, so R_4 is an **R**-algebra.

We conclude that *not every **R**-algebra is an **N**-algebra*, since in an **N**-algebra the star of an element is the supremum of all its powers. The example $\mathbf{R}\langle x \rangle$ shows that not every **N**-algebra is an **S**-algebra. We shall prove in this chapter that not every **A**-algebra is a **C**-algebra, and in the next, that not every **C**-algebra is an **R**-algebra.

Theorem 1. *An **R**-algebra satisfying the condition 'if $AB^n C \leqslant D$ for all n, then $AB^* C \leqslant D$' is an **N**-algebra.*

Sketch of proof. We define an *ideal* in the algebra as a set X of elements x such that $x \in X$, $y \in X$ and $z \leqslant x + y$ imply $z \in X$, and if $AB^n C \in X$ for all n, then $AB^* C \in X$. The *sum* of a collection of ideals is the least ideal containing them, and the product XY of two ideals is the least ideal containing every product xy ($x \in X$, $y \in Y$). We can prove (the details are rather subtle) that the ideals form an **S**-algebra, which contains the given algebra as the set of all *principal ideals*, defined as those ideals having maximal elements.

If $\mathbf{R}\langle x,y\rangle$ is the algebra of events in two letters, then the subalgebra of $\mathbf{R}\langle x,y\rangle \times R_4$ generated by $X = (x,2)$, $Y = (y,2)$ and $Z = (yx^*y,2)$ satisfies the condition 'if $AB^n \leqslant C$ for all n, then $AB^* \leqslant C$', but it is not an **N**-algebra, since $YX^n Y \leqslant Z$ for all n, but $YX^* Y \nleqslant Z$. This, and similar examples, suggests that the condition of Theorem 1 cannot be significantly weakened.

The propositions

$P0$: if $AB = BC$ then $A^* B = BC^*$

$P1l$: if $AB \leqslant BC$ then $A^* B \leqslant BC^*$ $P1r$: if $BC \leqslant AB$ then $BC^* \leqslant A^* B$

$P2l$: if $AB = B$ then $A^* B = B$ $P2r$: if $BC = B$ then $BC^* = B$

$P3l$: if $AB \leqslant B$ then $A^* B \leqslant B$ $P3r$: if $BC \leqslant B$ then $BC^* \leqslant B$

$P0^+, P1^+, P2^+, P3^+$: if $AB = BA$ then respectively

$$A^* B = BA^*, \qquad\qquad A^* B^* = B^* A^*,$$

$$A^* B^* = (AB)^* (A^* + B^*), \quad A^* B^* = (A + B)^*;$$

hold in all **N**-algebras. They fail in suitably chosen matrix algebras

over R_4 – for instance if $A = \begin{bmatrix} 1 & 1 \\ 0 & 2 \end{bmatrix}$ and $B = \begin{bmatrix} 2 & 0 \\ 2 & 2 \end{bmatrix}$, then

$AB = BA = \begin{bmatrix} 2 & 2 \\ 2 & 2 \end{bmatrix}$, but $A^* = \begin{bmatrix} 1 & 3 \\ 0 & 3 \end{bmatrix}$, and so $A^* B = \begin{bmatrix} 3 & 3 \\ 3 & 3 \end{bmatrix}$, while

$BA^* = \begin{bmatrix} 2 & 3 \\ 2 & 3 \end{bmatrix}$.

The proposition $P0$ is obviously equivalent to the conjunction of $P1l$ and $P1r$, and it is easy to see that $P1l$, $P2l$, $P3l$ are equivalent. I conjecture that $P0$ implies all propositions of this type (with finitely many hypotheses) – it certainly implies all the above. If the conjecture is true, then we have perhaps the nicest axiomatization of regular algebra, for then the axioms $P0$, $C1$–10 would not only imply all regular tautologies, but all of these 'propositions' as well.

Conservative algebras and the idempotency laws

A conservative event is a system of words with arbitrary cardinal number coefficients (or we can bound the cardinals by $\aleph_0$ to avoid certain logical problems). Axioms $C1$–12 and $C14$ hold for conservative events, but the idempotency laws $1^* = 1$, $A^{**} = A^*$, $1 + 1 = 1$, $A + A = A$, and $A^* A^* = A^*$ all fail. If we identify all finite non-zero coefficients we get a system satisfying $1 + 1 = 1$ and $A^* A^* = A^*$ but not $1^* = 1$.

The real numbers, with x^* defined as $1/(1 - x)$, satisfy $C1$–12 and $C14$, except that x^* is not always defined. But we cannot extend this partial algebra so as to overcome this, since for the reals we have $x^{***} = x$, while from $C1$–12 we can deduce $A^{****} = A^{***}$. But a

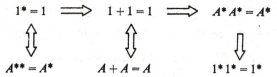

Fig. 12.2. Implications between the idempotency laws.

surprising amount of the rest of our theory holds for the reals, and in particular the matrix star formula $M1$ provides us with a rather peculiar algorithm for inverting real matrices! To prove $A^{****} = A^{***}$ we first observe that $1^* = 1 + 1^*$ by $C12$, and so $1^{**} = (1^* + 1)^* = 1^{**} 1^{***}$ by $C10$. But $1^{***} = 1 + 1^{**} 1^{***}$ by $C12$, so $1^{***} = 1 + 1^{**} = 1^{**}$. Now if $1 + A = A$, then we have $A^* = (1 + A)^* = A^* A^{**}$ by $C10$, and since $A^* A^{**} = 1 + A^* A^{**} = A^{**}$ by $C12$, this implies $A^* = A^{**}$. For any A we have $1 + A^{**} = A^{**}$, and so $A^{***} = A^{****}$. The free algebra on no generators defined by $C1$–12 is now easily constructed: its only elements are

$$0, 1, \quad 1 + 1 = 2, \ldots, \quad m(1^*)^n \, (m, n > 0), \quad \text{and} \quad 1^{**}$$

We can identify all the elements $m(1^*)^n$ to get an algebra satisfying $A^* A^* = A^*$ but not $1 + 1 = 1$. So we have all the implications of the diagram (Figure 12.2), but none but those indicated.

Acyclic algebras and the powerstar law C14

Using only $C1$–12 we can prove the equivalence of

$$(C14.m \text{ and } C14.n) \quad \text{with} \quad (C14.m \text{ and } C14.mn).$$

For assuming $C14.m$ we have

$$X^* = X^{n^*} X^{<n} \Rightarrow (X^m)^* = (X^m)^{n^*} (X^m)^{<n} \Rightarrow X^* = X^{mn^*}(X^m)^{<n} . X^{<m}$$

$$\Rightarrow X^* = X^{mn^*} X^{<(mn)} \Rightarrow X^* = (X^n)^{m^*} (X^n)^{<m} . X^{<n} = X^{n^*} X^{<n},$$

all equations being quantified over all X. If we adopt also $C13$, we can show that $C14.mn$ implies $C14.m$, and so that $C14.n$ is equivalent to the conjunction of $C14.p$ over all primes p not dividing n.

To see this, observe

$$X^* = X^{mn^*} X^{<(mn)} = X^{mn^*}(X^m)^{<n} . X^{<m} \leqslant X^{m^*} . X^{<m} \leqslant X^*,$$

assuming $C14.mn$. The last two steps use $C13$ in the assertion that for any function f we have $f(Y) \leqslant Y^*$.

We devote the next part of this chapter to the proof of a theorem of Redko to the effect that infinitely many of the axioms $C14.n$ are needed. We give two proofs, the first of which is the simple and elegant one given by Redko and Solomaa, but which unfortunately uses Redko's axiomatization of commutative events, and so would be considerably longer if given in full (and was hitherto incompletely proved). The second, and more cumbrous, proof, which is based on essentially the same idea, does not use this theorem, and is given as a prototype for some similar arguments which will be sketched in the next chapter.

Theorem 2. *For any finite set of regular tautologies, there is a prime p such that $C14.p$ is not deducible from the tautologies of that set, even supposed supplemented by all axioms $C14.n$ with $p\nmid n$.*

Proof. The given set of regular tautologies, since they hold for commutative regular events, must be deducible from $C1$–13, C^+1, C^+2, and $C14.n$ for a finite number of values of n. Choose a prime p which divides no one of these values of n. Observe that the axioms $C1$–13, C^+1, C^+2, and $C14.n$ for $p\nmid n$, have the property that if all the words in any starred part of one side have length divisible by p, the same is true of the other side, and that this property is preserved under taking consequences, but is not shared by $C14.p$.

We now proceed to the second proof, which since it does not rest on a finite axiomatic basis, must involve some general discussion of regular tautologies. Call a regular expression σ-free if it contains no occurrence of the symbol σ. Then the reader will have no difficulty in deducing, using only $C1$–13 that any regular expression can be reduced to a sum of terms each of which is *either* 0 or 1 or is simultaneously 0-free, 1-free, and +-free. The length of such a canonical expression is the number of non-zero terms, and the length of an equation between canonical expressions the maximal length of the sides.

Now using the axioms $C1$–13, and the supposition $x^p = 1$, it is easy to see that any regular expression in x can be reduced *either* to a sum of powers of x *or* to x^*, for if $y \geqslant x^i$ for some $i \not\equiv 0 \pmod{p}$ then $y^* \geqslant x^{ij}$ for all j, so $y^* \geqslant x$, so $y^* \geqslant x^*$, since for a suitable j we have $x^{ij} = x$. This observation, and the results

$$x^* x^* = x^{**} = x^i x^* = x^* x^i = x^* + x^i = x^*$$

do indeed show that the 2^p formal sums of powers of x, together with the formal symbol x^*, satisfy the axioms $C1$–13, and we can check that they also satisfy $C14.n$ if $p \nmid n$. They therefore form an **A**-algebra A_p (say) which is *not* a **C**-algebra, since $C14.p$ fails at x.

Theorem 3. *Any regular tautology of length less than p when both sides are canonical expressions, holds in A_p.*

Proof. We consider a minimal tautology $f(A, B, \ldots) = g(A, B, \ldots)$ which fails in A_p, say $f(A_0, B_0, \ldots) \neq g(A_0, B_0, \ldots)$, for some elements $A_0, B_0, \ldots$ of A_p. If any of $A_0, B_0, \ldots$ are 0 or 1, we can replace the corresponding A or B or . . . by 0 or 1 and get a simpler tautology which fails. Since A_p becomes an **R**-algebra when we identify x^* with $x^{<p}$, we can suppose $f(A_0, B_0, \ldots) = x^{<p}$, $g(A_0, B_0, \ldots) = x^*$. Now the variables among $A, B, \ldots$ which appear inside stars on *either* side of the tautology must appear within the stars on *both*, since these are the variables which appear arbitrarily often in suitably chosen words of the regular event $f(A, B, \ldots) = g(A, B, \ldots)$. For any such variable X the corresponding X_0 must be a single power x^i of x, for otherwise the star containing X_0 on the left would have value x^*, as therefore would $f(A_0, B_0, \ldots)$. If we replace the remaining variables Y, not appearing within the stars on either side, by 1, the value of the left-hand side cannot be changed *into* x^*, while the value of the right-hand side cannot be altered *from* x^*, since some starred expression on the right must have value x^*. The tautology therefore still fails, but now each

term on the left-hand side has value a single power of x, and so there must be at least p of them, a contradiction.

So no finite system of **R**-tautologies is a sufficient system of axioms for regular algebra, or even for the weaker notion of **C**-algebra. On the other hand, it is natural to conjecture, and several authors have done so, that there is a finite system of axioms for regular algebra which becomes complete when we add the axiom-scheme $C14$. In every case the proposed system has been equivalent to $C1$–14. It is therefore of interest that this system is *not* complete, and the conjecture is false even in the more general form above, for we shall prove in the next chapter that any complete axiom system needs infinitely many axioms involving two or more variables.

If we allow additional rules of proof such as our propositions $P0$, $P1$, etc., then we can make do with finitely many axioms.

Theorem 4. (*Solomaa's axiomatization.*) *The system $C1$–13, supplemented by the rule 'if $G = F + EG$ is a tautology for regular expressions E, F, G such that $E \geqslant 1$ is not a tautology, then $G = E^* F$ is a tautology', is a complete axiom-system for regular algebra.*

Proof (in outline). Take a regular expression E in inputs $a, b, c, \ldots$ we show by differentiation that we can produce a finite number of regular expressions corresponding to the word derivates of E, such that for any one of these expressions F, say, there are others, $G, H, \ldots$, say, for which

$$F = (0 \text{ or } 1) + aG + bH + \ldots$$

is deducible from $C1$–14. We call this system of equations the system corresponding to the machine obtained by differentiating E. If we consider two expressions E_1 and E_2 which represent the same event, we get in this way two systems corresponding to two machines M_1 and M_2 representing this event, each variable of either system corresponding to a state of the appropriate machine.

Now *both* systems of expressions, suitably repeated, satisfy the equations corresponding to the accessible part of direct product machine $M_1 \times M_2$, the variable corresponding to the state (F_1, F_2) of this being F_1, in the first system, and F_2 in the second. But using

Solomaa's rule we can 'solve' any such system of equations piecemeal, showing that it has a unique solution, and hence that $E_1 = E_2$. The idea is essentially that the Solomaa rule implies itself for matrices.

The rather awkward phrasing of the theorem can be removed at the cost of some complication in the proof. We assert that the classical axioms, supplemented by our rule $P2$: 'if $AB = B$ is a tautology, then so is $A*B = B$', also form a complete system. We can rephrase this in the form

Theorem 5. *Any* **A***-algebra for which* $A*B = B$ *whenever* $AB = B$, *is an* **R***-algebra.*

We do not prove this theorem. There does not seem to be a similar way of expressing Solomaa's theorem in terms of algebras. We might add, that in both Theorems 4 and 5 the last few of the classical axioms are redundant, being derivable from the rules.

Yanov has given a complete system of axioms for events $\geqslant 1$, namely

1   $X + X = X$
2   $X + Y = Y + X$
3   $(X + Y) + Z = X + (Y + Z)$
4   $(XY)Z = X(YZ)$
5   $X(Y + Z) = XY + XZ$
6   $(X + Y)Z = XZ + YZ$
7   $X^{**} = X^*$
8   $X^* X^* = X^*$
9   $(X + Y)^* + X = (X + Y)^*$
10   $(X + Y)^* = (X^* + Y)^*$
11   $(X + Y)^* = (XY)^*$
12   $(XY)^* X = (XY)^*$
13   $X(XY)^* = (XY)^*$
14   $XY + Y = XY$
15   $XY + X = XY$.

It is easy to check that each of these is deducible from $C1-14$ together with the additional assumptions $X \geqslant 1, Y \geqslant 1, Z \geqslant 1$. So in this sense $C1-14$ can be said to be complete for events $\geqslant 1$.

CHAPTER THIRTEEN

The strength of the classical axioms

We have already been forced to sketch certain proofs, leaving details to the reader. We now warn him that this policy will be adopted wholesale in this chapter, where complete proofs would be intolerably tedious. Our aim is to describe the conditions under which we should expect a regular tautology to be deducible from $C1$–14, and in so doing, to lay the foundation for a possible completion of this axioms-system.

Theorem 1. *In $\mathbf{C}\langle x, y, \ldots \rangle$, the free $\mathbf{C}$-algebra on n letters, it is possible in a unique way to define input derivates E_x, E_y, $\ldots$ of an expression E so that $E = o[E] + xE_x + yE_y + \ldots$.*

Proof. The elements of $\mathbf{C}\langle x, y, \ldots \rangle$ are, of course, equivalence classes of regular expressions under *classical equivalence*. This relation means that equality is deducible from $C1$–14 and can be formally defined as the least equivalence relation $\sim$ for which $A \sim B$, $C \sim D$ implies $A + C \sim B + D$, $AC \sim BD$, and $A^* \sim B^*$, and such that $L \sim R$ whenever A, B, C are regular expressions and L, R the two sides of one of the axioms $C1$–14.

Now, for E a regular expression we *define E_x, E_y, $\ldots$, $o[E]$* by $D1$–9 of Chapter 5, and show the definitions valid. This entails showing that $L_x = R_x$ whenever L and R are as above. For instance, with $C10$:

$$L_x = (A_x + B_x)(A + B)^*,$$
$$R_x = A_x A^* + B_x (A^* B)^* A^* + A_x A^* B . (A^* B)^* A^*,$$

and the equality is fairly easy to prove.

We now know that the classical equivalence of E and F implies that of E_x and F_x. The uniqueness therefore follows, for if $H, K, \ldots$ are any expressions with $E = (0 \text{ or } 1) + xH + yK + \ldots$, we deduce $H = E_x$, $K = E_y$, $\ldots$ on differentiating.

Corollary. *If* $Af(A,B,\ldots) = Ag(A,B,\ldots)$ *is a* **C**-*tautology, then so is* $f(A,B,\ldots) = g(A,B,\ldots)$.

We shall use the Corollary implicitly later. From now on, we work inside some **C**-algebra, and make deductions using only $C1$–14.

We define the star of a matrix inductively by the matrix star formula

$M1$ of Chapter 3, the matrix $M = \begin{bmatrix} A & B \\ C & D \end{bmatrix}$ being partitioned so that

A and D are square.

Theorem 2. *Stars of matrices are well-defined. The axioms* $C1$–13 *imply themselves for matrices.*

Proof. We prove these together by induction on the order of the matrices. For 2×2 matrices there is no problem about well-definedness, and each of the axioms reduces to four tautologies in, at most, twelve variables, whose verification we leave to the reader (most of them are easy!). The same proof will work for larger matrices given the inductive assumptions and the well-definedness of the stars. The latter reduces to showing that when we compute the star of the matrix

$$\begin{bmatrix} A & B & C \\ \hline D & E & F \\ G & H & I \end{bmatrix} = \begin{bmatrix} A & B & C \\ D & E & F \\ \hline G & H & I \end{bmatrix}$$

in the two indicated ways we get the same result, and this is just the verification of nine tautologies in nine variables each.

$C13$ (which can be taken in the form $C13°$) is only used in proving its own matrix version. This happy state of affairs will not persist! A similar proof can be given to show that $C14.2$ implies itself for matrices, but I have never been able to verify directly that $C14.3$ implies itself for 2×2 matrices.

We can now show that reordering of the rows and columns (in the same way) has the expected effect. To see this, observe that the reordered matrix is $\pi^{-1} M \pi$ for a permutation matrix π, and that

$$(\pi^{-1} M \pi)^* = 1 + \pi^{-1}(M \pi \pi^{-1})^* M \pi = \pi^{-1}(1 + M^* M) \pi = \pi^{-1} M^* \pi,$$

using $C12$ and the equations $\pi \pi^{-1} = \pi^{-1} \pi = 1$ (π^{-1} is just the transpose of π).

Now let $C14.n^*$ denote the tautology in n letters $x_0, \ldots, x_{n-1}$ obtained as follows. Define expressions $E_0, E_1, \ldots, E_{n-1}$ in these letters by

$$\begin{bmatrix} x_0 & x_1 & x_2 & \ldots & x_{n-1} \\ x_{n-1} & x_0 & x_1 & x_2 & \\ & & x_0 & x_1 & \ldots \\ x_1 & & & & x_0 \end{bmatrix}^* = \begin{bmatrix} E_0 & E_1 & E_2 & \ldots & E_{n-1} \\ E_{n-1} & E_0 & E_1 & E_2 & \ldots \\ & & E_0 & E_1 & \ldots \\ E_1 & & \ldots & & E_0 \end{bmatrix}.$$

Then $C14.n^*$ is the tautology $E_0 + E_1 + \ldots + E_{n-1} = (x_0 + \ldots + x_{n-1})^*$.

Theorem 3. $C14.n^*$ *implies* $C14.n$ *for matrices.*

Proof. As in Theorem 2, it will suffice to consider the case of 2×2 matrices. Now let $X = \begin{bmatrix} a & b \\ c & d \end{bmatrix}$ be a 2×2 matrix, and consider the $2n \times 2n$ matrix

$$M = \begin{bmatrix} 0 & X & 0 & \ldots & 0 \\ 0 & 0 & X & \ldots & 0 \\ & & \ldots & & \\ 0 & 0 & \ldots & 0 & X \\ X & 0 & \ldots & 0 & 0 \end{bmatrix}.$$

Partitioning M as indicated, we readily find that M^* is a circulant matrix in 2×2 blocks, the top row of blocks being X^{n^*}, $X^{n^*+1}, \ldots,$ X^{n^*+n-1}, and so

$$\begin{bmatrix} 1 & 0 & 0 & 0 & \ldots \\ 0 & 1 & 0 & 0 & \ldots \end{bmatrix} . M^* . \begin{bmatrix} 1 & 0 \\ 0 & 1 \\ 1 & 0 \\ 0 & 1 \\ & \vdots \end{bmatrix} = X^{n^*} X^{<n}.$$

It therefore remains to be shown that the displayed product is X^*. But we can permute the rows and columns to reduce this to proving that

$$Y = \begin{bmatrix} 1 & 0 & 0 & 0 & 0 & 0 & 0 & 0 \\ 0 & 0 & 0 & 0 & 1 & 0 & 0 & 0 \end{bmatrix} \begin{bmatrix} a & & & & b & & & \\ & a & & & & b & & \\ & & a & & & & b & \\ & & & a & & & & b \\ \hline c & & & & d & & & \\ & c & & & & d & & \\ & & c & & & & d & \\ & & & c & & d & & \end{bmatrix}^* \begin{bmatrix} 1 & 0 \\ 1 & 0 \\ 1 & 0 \\ 1 & 0 \\ \hline 0 & 1 \\ 0 & 1 \\ 0 & 1 \\ 0 & 1 \end{bmatrix} = X^*,$$

(where for convenience we have illustrated the case $n = 4$). Now the leading block of the star of the displayed square matrix is $(A + BD^*C)^*$, where A, B, C, D are the component $n \times n$ matrices, and the matrix $A + BD^*C$ is readily computed as the circulant matrix

$$\begin{bmatrix} X_{n-2} & X_{n-1} & X_0 & X_1 & \cdots \\ & X_{n-2} & X_{n-1} & X_0 & \cdots \\ & & \cdots & & \\ X_{n-1} & X_0 & & \cdots & \end{bmatrix}$$

where X_i is $bd^{n^*+i}c$, except that $X_{n-1} = a + bd^{n^*+n-1}c$. The leading term of Y is the sum of the terms in the top row of this circulant, and so we must prove that this is the leading term of X^*, namely $(a + bd^*c)^*$. (The other terms reduce to the same problem.) But the sum of the terms X_i in the top row of the above circulant is $a + bd^{n^*}d^{<n}c = a + bd^*c$. So our problem is to prove that the star and the sum of the terms of the top row of the above matrix is the sum of the terms in the top row of its star, and this is precisely what $C14.n^*$ guarantees.

Now we shall reduce $C14.n^*$ to the following proposition, called $P(n)$.

'If $E_0, E_1, \ldots, E_{n-1}$ are expressions such that

(i) $E_i E_j \leqslant E_{i+j}$ (subscripts added $\mod n$),
(ii) $E_0^* = E_0$,

then $(E_0 + \ldots + E_{n-1})^* = E_0 + \ldots + E_{n-1}$.'

Theorem 4. $P(n)$ *implies* $C14.n^*$.

Proof. The functions E_i appearing in $C14 . n^*$ are the same functions of the E_j as they are of the x_j. This follows from the formula $M^{**} = M^*$ for matrices, which we now know to hold. So, since we can prove $x_j \leqslant E_j$, we get the 'promotion lemma' – if for some regular function f we can prove $f(x_0, \ldots, x_{n-1}) \leqslant E_i$, then we can deduce $f(E_0, \ldots, E_{n-1}) \leqslant E_j$. It is fairly easy to deduce $x_0^* \leqslant E_0$ and $x_i x_j \leqslant E_{i+j}$, and so we can deduce that the hypotheses of $P(n)$ hold for the E_i. The conclusion is that $(E_0 + \ldots + E_{n-1})^* = E_0 + \ldots + E_{n-1}$, implying $(x_0 + \ldots + x_{n-1})^* \leqslant E_0 + \ldots + E_{n-1}$. On the other hand, we also have the reverse inequality, since the E_i are functions of the x_j, and putting both together we get the conclusion of $C14 . n^*$.

It might be amusing to see the form of the E_i for small i. Writing $x, y, z, \ldots$ for $x_0, x_1, x_2, \ldots$, we have, for $n = 2$

$$E_0 = (x + yx^* y)^*, \qquad E_1 = E_0 . yx^*,$$

and for $n = 3$

$$E_0 = \{x + y(x^* zx^* y)^* (x^* z + x^* yx^* y)$$
$$+ z(x^* zx^* y)^* (x^* y + x^* zx^* y)\}^*$$

and

$$E_1 = E_0 . \{zx^* z(x^* yx^* y)^* x^* + y(x^* zx^* y)^* x^*\}.$$

The boiling process

We now intend to deduce $P(n)$ from $C1$–14. We introduce the abbreviations $E_{ijk+lmpq+}\ldots$ for $E_i E_j E_k + E_i E_m E_p E_q + \ldots$ (typically), and $A \leqslant B$ to mean that $XA^* X \leqslant XB^* X$, where $X = E_{0+1+\cdots+(n-1)}$. To prove $P(n)$ it suffices to show that $X \leqslant E_1$, for this gives $X^* \leqslant X(E_1)^* X \leqslant XXX \leqslant X$, since $(E_1)^* = (E_1)^{n*}(E_1)^{<n}$, and $E_1^n \leqslant E_0$, so that we get $E_1^* \leqslant X$.

Lemma. (i) $E_0 + S \leqslant S$ if S is of the form $E_{ijk+}\ldots$.

(ii) $S + T \leqslant S$ if $T \leqslant$ some product of terms of S.

(iii) $E_{ij} \leqslant E_{(i+j)}$.

(iv) $S + T \leqslant T^{\leqslant n} S$, if T is a single term $E_i E_j E_k \ldots$ and $T^{\leqslant n}$ denotes $E_0 + T + \ldots + T^{n-1}$.

Note. In (iv) we could also use $ST^{\leqslant n}$. Either is called a *boiling operation*.

Proof. In (i) we use $(E_0 + S)^* = E_0^*(SE_0^*)^*$ and the facts that $E_0^* \leqslant X$ and $E_i E_0^* = E_i E_0 \leqslant E_i$. In (ii) we have $T \leqslant S^*$, so $(S + T)^* = S^*$. (iii) is obvious, and for (iv) we have

$$(S + T)^* = (T^* S)^* T^* = T^{n*} T^{<n}(ST^{n*} T^{<n})^*.$$

In this, using $T^n \leqslant E_0$, we can replace $T^{n*} T^{<n}$ by $E_0^* T^{<n} \leqslant T^{\leqslant n}$, and this, together with $T^{\leqslant n} \leqslant X$, proves the result.

An example of the boiling process

We take $n = 6$, and boil down the expression $E_{0+1+2+3+4+5}$ as follows:

$E_{(0)+1+2+3+4+5}$

$\leqslant E_{1+2+3+4+(51)+(52)+(53)+(54)+(55)(1)+\ldots+(55555)(4)}$

$\leqslant E_{1+2+3+4}$

$\leqslant E_{1+2+3+41+(42)+(43)+\ldots+(44444)(3)}$

$\leqslant E_{1+2+3+41}$

$\leqslant E_{1+2+3+(411)+(412)+(413)+(41411)+\ldots}$

$\leqslant E_{1+2+3}$

$\leqslant E_{1+2+31+32+(331)+\ldots}$

$\leqslant E_{1+2+31+(321)+\ldots}$

$\leqslant E_{1+2+311+(312)+\ldots}$

$\leqslant E_{1+2+(3111)+\ldots}$

$\leqslant E_{1+21+221+(2221)+\ldots}$

$\leqslant E_{1+21+(2211)+\ldots}$

$\leqslant E_{1+211+(21211)+\ldots}$

$\leqslant E_{1+2111+(21111)+\ldots}$

$\leqslant E_{1+(21111)+\ldots}$

$\leqslant E_1.$

Here we have bracketed any term which may be struck out using rules (i), (ii) and (iii). The second line is obtained from the first by boiling with E_5 for T. The term $E_{555554} = E_{55555}E_4 \leqslant E_1 E_4$, so can be struck out by rule (ii), and similar observations show that we can strike out the remaining bracketed terms, obtaining the third line, which yields the fourth on boiling by $T = E_4$, reducing to the fifth, which boils to the sixth by E_{41}, and so on.

Why does the boiling process terminate?

Theorem 5. *Suppose, starting with $E_{1+2+\ldots+(n-1)}$ we always boil with the last term, ordering the terms so obtained as*

$$S_1 + S_2 + \ldots + S_k + TS_1 + \ldots + TS_k + T^2 S_1 + \ldots + T^{n-1} S_k,$$

and then cancel all the terms we can. Then the boiling process always terminates, and we have at every stage

(i) *the length $i + j + \ldots$ of any remaining term $E_{ij} \ldots$ is $<n$;*
(ii) *if any term is 'left-truncated' the result is $\leqslant$ some product of earlier terms;*
(iii) *the sum $\sum 2^{-\text{length}}$ over all terms strictly decreases as we boil;*
(iv) *as we move right along the expression the lengths of the terms are strictly increasing.*

Proof. A *left-truncation* of any term $E_{ij\ldots klm\ldots}$ is any term such as $E_{k'lm\ldots}$, where $k' \leqslant k$, other than the term itself. So, for instance, E_{345} has truncations

$$E_{245}, E_{145}, E_{45}, E_{35}, E_{25}, E_{15}, E_5, E_4, E_3, E_2, E_1, E_0.$$

Now property (iv) is obviously preserved from stage to stage, since we always boil with the largest term, and property (iii) then becomes obvious when we consider the sum as a binary number. To prove (ii), we observe that any truncation of $T^q S_i$ (say) is *either* a truncation of S_i, or of the form $T' . T^p S_i$, where $p < q$ and T' is a truncation of T. In the first case it is $\leqslant$ a product of terms S_j ($j < i$) by the property applied to the expression before boiling by T. In the second case, it is a product of terms T', T, S_i, and T' is a product of the S_j by the property before boiling. So (ii) is preserved under boiling.

To prove (i), suppose the term $E_{ijk\ldots}$ has length $\geqslant n$. Then we can amalgamate just enough of the subscripts $i, j, k, \ldots$, starting at the left, to have sum $\geqslant n$, when we obtain a left-truncation of the term, which can therefore be struck out as being $\leqslant$ some product of earlier terms.

Corollary (*at last!*). $C14.n$ *implies itself for matrices, in the presence of $C1$–13.*

Note that $C13$ is used very heavily. I do not know how to prove the corresponding result (which is surely true) without using $C13$.

The above investigation suggests the following ideas. We define for each finite semigroup G with unit 1, a proposition $P(G)$ as follows.

'If we have expressions E_g $(g \in G)$ such that

(i) $E_g E_h \leqslant E_{(gh)}$,
(ii) $(E_{g,h})^* = E_{g,h}$, where $E_{g,h}$ is the sum $\sum E_k$ $(gk = h)$

then $E_G^* = E_G$, E_K denoting generally $\sum E_k$ $(k \in K)$.'

I can prove fairly easily that this *proposition $P(G)$* is equivalent to a *tautology $R(G)$*, defined as follows. Let x_g $(g \in G)$ be distinct letters, and define $x_{g,h}$ as the sum $\sum x_k$ over k such that $gk = h$. Then define expressions $E_{g,h}$ in these letters by requiring that the matrix of expressions $E_{g,h}$ be the star of the matrix of sums $x_{g,h}$. Then $R(G)$ is the tautology $(\sum x_g)^* = \sum E_g$, both sums over all $g \in G$.

It seems likely that one can prove that $R(G)$ for all groups implies $R(G)$ for all semigroups. It also seems possible to prove that $R(G)$ for all semigroups implies all regular tautologies. (To deduce the tautology $E = F$ we apply the $R(G)$ for which G is the direct product of the semigroups of state maps corresponding to machines representing E and F, classically. A machine with corresponding matrices L, M, N *classically represents E* if $E = LM^*N$ as a consequence of $C1$–14. Each input defines a map on the states of this machine, and these maps generate the semigroups mentioned.) However, I have concentrated on the problems concerned in simplifying the axioms $R(G)$, supposing them, as seems very likely to be provable, to be a complete basis for the regular tautologies in the presence of $C1$–14. I consider here only the equivalent form $P(G)$.

Theorem 6. *If G is a soluble group, then $P(G)$ is deducible from $C1$–14.*

Proof. We consider expressions E_g $(g \in G)$ as in the definition of $P(G)$, and define E_K as $\sum E_k$ $(k \in K)$ for any subset K of G. Since G is soluble, we can select a normal subgroup H and an element g of G, in such a way that G is the disjoint union $H \cup Hg \cup \ldots \cup Hg^{n-1}$, and $g^n \in H$. Then the elements $E_i = E_{Hg^i}$ satisfy the conditions of $P(n)$, if we inductively suppose ourselves to have proved $P(H)$, and the conclusion of $P(n)$ in this case proves $P(G)$.

Indeed $P(n)$ is $P(G)$ for a cyclic group G of order n. Our proof of Theorem 6 really shows that $P(G)$ is implied by $P(H)$ as H runs through the composition factors of G, so that it suffices to prove $P(G)$ for all simple groups G. These considerations suggest that perhaps the classical axioms do not imply $P(G)$ for insoluble G, and that in some sense a natural set of axioms would be $C1$–13 supplemented by $P(G)$ for all simple G, with $C14.p$ corresponding to the case when G is cyclic of prime order p. We now prove that indeed for insoluble G, $P(G)$ is not derivable from $C1$–14.

We consider an insoluble group G, with a family F of subgroups containing $\{1\}$ and containing, whenever it contains K, any group H in which K is normal with cyclic quotient. We define G_∞ as the semigroup $G \cup \{\infty\}$, with $x\infty = \infty x = \infty\infty = \infty$ for all $x \in G$. Then we construct an algebra $\mathbf{C}(G) = \mathbf{C}_F(G)$, as follows. The elements of $\mathbf{C}(G)$ are the subsets of G_∞, with the usual addition and multiplication. The usual definition of A^* would be the semigroup $\langle A \rangle$ generated by A. This is either a group, or a group together with ∞. We instead define A^* as $\langle A \rangle$, when $A \in F$, and as $\langle A \rangle \cup \{\infty\}$ in all other cases. For the family F, we might take the set of all soluble groups inside G; this shows that F need not include G.

Theorem 7. $\mathbf{C}(G)$ *is a* $\mathbf{C}$*-algebra for each insoluble G.*

Proof. Axioms $C1$–10 and $C13^0$ are immediate. For $C11$, we certainly have $(A + B)^* = (A^*B)^*A^*$ if $\infty \in A + B$, so that it suffices to show that $\langle A + B \rangle$ is insoluble if and only if $\langle\langle A \rangle B\rangle$ is insoluble, A and B being subsets of G, $B \neq 0$. But then we have $\langle A + B \rangle = \langle\langle A \rangle B\rangle$, since every $a \in A$ can be written in the form $ab.b^k$ for some $b \in B$. For $C12$, it suffices to prove $\langle AB \rangle = \langle BA \rangle$ for subsets A and B of G, and this is just as easy. For $C14.n$, all will be well unless $\langle A \rangle \notin F$, while $\langle A \rangle \in F$. But since it is easy to see that $\langle A^n \rangle$ is a normal subgroup of $\langle A \rangle$ with cyclic quotient, this cannot happen.

On the other hand, the proposition $P(G)$ fails in $\mathbf{C}(G)$ if $G \notin F$, for taking $E_g = g$ for each $g \in G$ it would yield $G^* = G$, whereas for $G \notin F$ we have $G^* = G + \infty \neq G$. So using the equivalence of $P(G)$ with $R(G)$, we deduce that *the classical axioms are not sufficient to prove* *the tautology R(G) for any insoluble G.* So that the reader can see a full proof of this, we improve it to the following:

Theorem 8. *Let Rn denote the axiom*

$$(A + B)^* \quad = [(A + B)\{B + (AB^*)^{n-2} A\}] \times$$
$$\times [1 + (A + B)\{1 + AB^* + \ldots + (AB^*)^{n-2}\}].$$

Then Rn is not derivable from C1–14 if $n \geqslant 5$.

Proof. The reader should find Fig. 13a helpful in verifying that *Rn* is indeed a tautology, and in following some of the argument. We take $n = 5$ for simplicity (in two senses!), let $G = \sum_5$, the symmetric group on 5 letters, and $a = (12345)$, $b = (12)$, two generating elements of *G*. We easily check that every element of the subgroup

$$\langle\langle a + b\rangle(b + (a\langle b\rangle)^3 a)\rangle$$

fixes 1, and since the subgroup of *G* fixing 1 is $\sum_4$, a soluble group, we have $\infty \notin [(a + b)\{b + (ab^*)^3 a\}]$, and we rapidly deduce

$$\infty \notin [(a + b)\{b + (ab^*)^3 a\}][1 + (a + b)\{1 + ab^* + (ab^*)^2 + (ab^*)^3\}]$$

proving that *R5* fails in **C**(G) if *F* is the family of soluble subgroups of *G*, since $\infty \in (a + b)^*$, $\sum_5$ being insoluble. For $\sum_n$ we take for *F* the family of all subgroups other than A_n or $\sum_n$. The argument fails if $n \leqslant 4$, and indeed *Rn* is deducible from *C1–14* in these cases, although the proofs are very long indeed.

It is possible to prove along the lines of our proof of Theorem 3 of the last chapter, that every finite set of regular tautologies holds in some **C**($\sum_n$) with the above choice of *F*, while *Rn* does not. Since also *C1–14* hold in **C**($\sum_n$), so does every 1-variable regular tautology. So we conclude

Theorem 9. *Any complete axiom system for the set of regular tautologies must involve infinitely many axioms involving two or more variables.*

I conjecture that the classical axioms together with the axioms *Rn* form a complete system, and the previous remarks have already suggested a possible line of proof. *Rn* is actually obtained from the tautology $R(\sum_n)$ in *n*! variables by putting all but two of the variables equal to zero, and perhaps some kind of boiling operation can be used to reduce the number of variables in $R(\sum_n)$ to these – one can certainly boil some of them away. Of course $R(\sum_n)$ implies $R(G)$ for all finite groups G. If we are prepared for axioms involving rather more

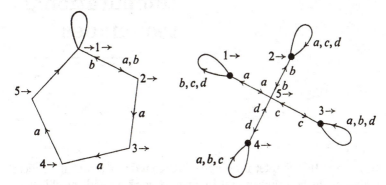

Fig. 13.1. The tautology R5.   Fig. 13.2. The tautology R′5.

variables, we get a rather more elegant conjecture: defining tautologies $R'n$ in $n-1$ variables $A, B, C, \ldots$ by

$$R'n: (A + B + \ldots)^* = [A\bar{A}^* A + B\bar{B}^* B + \ldots]^*$$
$$\times [1 + A\bar{A}^* + B\bar{B}^* + \ldots]$$

where $\bar{A} = B + C + \ldots$, $\bar{B} = A + C + \ldots$, etc., we can prove (by setting $a = (1n)$, $b = (2n)$, $\ldots$) that $R'n$ is not classical if $n \geqslant 5$, and we conjecture that the tautologies $R'n$ also supplement C1–14 to a complete set. $R'n$, like Rn, is deducible if $n \leqslant 4$, but even for $n = 3$ it is difficult to produce a proof without using the general ideas of this chapter, and for $n = 4$ I doubt if a completely written out proof could be fitted into 10 pages.

CHAPTER FOURTEEN

Some computational techniques

In a sense, this chapter is an appendix, containing a ragbag of ideas which might be found useful by anyone working with machines or regular expressions. We have in mind manual rather than automatic computation, and have added a number of exercises of varying difficulty which we advise the reader to attempt. Some of the solutions will be found after Chapter 15.

Manipulating regular expressions

Use inequalities wherever possible. In 'naturally given' expressions, it often happens that one term is absorbed in another, frequently by use of the formulae

$$A + (B + C + \ldots)^* = (B + C + \ldots)^* = (1 + A)(B + C + \ldots)^*$$

which hold whenever $A \leqslant f(B, C, \ldots)$. If the factors in a term of form $\ldots A(BC\ldots A)^* BC \ldots$ recur periodically, and the starred bracket contains an exact period, then this bracket can be moved left or right (using $(XY)^* X = X(YX)^*$) in any way so as still to contain a period. If an extreme factor of the bracket is starred, we can often use the tautologies

$$(X^* Y)^* X^* = (X + Y)^* = (Y^* X)^* Y^*,$$

$$(X^* Y)^* = 1 + (X + Y)^* Y$$

to simplify further.

Very complicated regular expressions are often handled easily in matrix form. In the last resort, of course, we test regular expressions for equality by building machines – a particularly simple technique will soon be described.

Exercises. (To be derived from $C1$–14.)

1    $(xy^* + yx)^* = (yx)^* + (yx)^* x(x + y)^*$.
2    $(1 + x^* y)(x + yx^* y)^* = (x + y)^*$.
3    $(xy)^* [x + xy(yy^* x)^*]^* = [(xy^* y)^* yx + x]^* (xy)^*$.
4    $(x + y)^* = (xy^* x + yx^* y)^* (1 + xy^* + yx^*)$ ($R'3$ of Chapter 13).
5   R' 4 of Chapter 13.

Constructing machines from regular expressions

Use repeated input derivation as in Chapter 5, but simplifying the resulting expressions whenever possible. When one derivate is a sum of others, most of the work of computing its derivates will already be done, and indeed if we are satisfied with a linear mechanism (say) instead of an orthodox machine, we need not include a node for any such derivate (this often brings the number of nodes down from N to about $\log N$). In simple cases the process usually produces the Moore reduced form, especially if we keep an eye on inequalities between the derivates as we find them. If not, the commonest reductions are to identify all states of any system closed under the inputs and having the same output, or all states of any system on which the output and all the input functions agree.

Exercises

6   Find a machine for $(y^2 x + xy^* xy)^* (x^2 + yx^*)^*$.
7   Find machines for other expressions of comparable size. (Your *expertise* is the ratio
     (size of reduced machine)/(size of machine first obtained).
This should usually be about 90% for machines of about 20 states.)

Constructing regular expressions from machines

Quite inexperienced workers can often write down at sight regular expressions for simple machines, and on occasion such expressions will be correct! The method is usually to calculate the effects of individual parts of the machine and then fit these together piece by piece, and we can formalize it in terms of general mechanisms (Chapter 5) – for instance $a \underset{\longrightarrow}{\overset{b}{\bigcirc}} c$ becomes $\underline{a\ b^* c}$ In matrix terms we are simplifying by replacing the machine transition matrix by smaller matrices with more general entries.

The best general approach is apparently to make as many such simplifications as we easily can, and then to write down a system of one-sided linear equations (as in Chapter 10, but the idea is already in Chapter 3) whose solutions include the required expressions, and solve these piecemeal.

The simplest expressions for the star of a 3×3 matrix are probably those implied by

$$J = [(A + CI^* G) + (B + CI^* H)(E + FI^* H)^* (D + FI^* G)]^*$$
$$K = J.(B + CI^* H)(E + FI^* H)^*$$

for the first two terms in the star of

$$\begin{bmatrix} A & B & C \\ D & E & F \\ G & H & I \end{bmatrix}.$$

The general term in the star of an upper triangular matrix is obtained by 'bouncing down the diagonal' in all possible ways. Thus the unwritten term in

$$\begin{bmatrix} A & B & C & D \\ & E & F & G \\ & & H & I \\ & & & J \end{bmatrix}^* =$$

$$\begin{bmatrix} A^* & A^* BE^* & A^* BE^* FH^* + A^* CH^* & \cdots \\ & E^* & E^* FH^* & E^* FH^* IJ^* + E^* GJ^* \\ & & H^* & H^* IJ^* \\ & & & J^* \end{bmatrix}$$

is $A^* BE^* FH^* IJ^* + A^* CH^* IJ^* + A^* BE^* GJ^* + A^* DJ^*$.

The exercises for this section and the next are given together.

Boolean functions

To compute a regular expression for, say, $E \backslash F$, we can build a machine for $E \backslash F$ by repeated input derivation, using $(E \backslash F)_w = E_w \backslash F_w$, and then compute an expression for this machine. We can simplify the process by stopping the derivation at any state for which we already know a regular expression, for instance when we can see that $E_w \leqslant F_w$, or that $E_w \backslash F_w = 0$, when we can write down 0 or E_w respectively. The resulting hybrid, partly a machine and partly a collection of undifferentiated regular expressions, can then be regarded as a general mechanism, and a regular expression deduced using the above ideas.

Exercises

8   Find a regular expression for $(xy^* + yx)^* \cap (y^* x + xy)^*$.

9   Prove that $(xy^* + yx)^* \cap (y^* x + xy)^* = (yx)^* [x + xy(yy^* x)^*]^*$,

Testing inclusions and equalities

A simple test for the inclusion $E \leqslant F$ is to build a machine for $E \backslash F$ and check that no state emits an output. Usually only one or two differentiations are needed even for quite complicated expressions, since almost all the derivates can be rejected on trivial grounds. It is often suggested that we test two expressions for equality by building *two* machines and putting these into reduced form – of course, we can instead build a *single* machine for the symmetric difference $E \varDelta F = (E \backslash F) + (F \backslash E)$. Again, in the typical case most of the differentiations can be avoided.

Exercise

10   Check Exercises 1–5 by this method.

Calculations with factors

The best advice is – avoid such calculations. If we really must compute factor matrices, we can regard the factors as intersections of word derivates, and find expressions for large numbers of them simultaneously. But, for instance, to find best approximations and approximating functions we need compute only the left and right factors and then the best (constant + linear)-approximation to E_{ij} by the F_t is the sum of all F_t with $L_i.F_t.E_j \leqslant E$, adding 1 if $L_i R_j \leqslant E$. We can do better by selecting suitable words u_i and v_j and finding the sums of the F_t with $u_i F_t v_j \leqslant E$, which sums contain all the required information if the u_i, v_j are cleverly chosen. (What we need is that every left word derivate of E is a $\partial E/\partial u_i$, and every right word derivate a $(\partial E/\partial v_j)r$.)

Often in the test we have used factors, where a less inclusive system suffices. Usually we use only that the factorizations $L_i.R_i$ (or $E_{pi}E_{ij}...E_{qr}$) dominate every subfactorization of E into words $u.v$ (or $u.v...w$).

The solution of equations in general

The factor method for solving certain equations is usually inferior to *ad hoc* techniques, unless general theorems are required. For other types of equation we offer the following remarks.

The starth root of E (Chapter 6) is $(E\backslash 1)\backslash(E\backslash 1)^{2+*}$, and for the occasional practical application there is a simple machine construction based on this formula.

Many equations can be written in operator form, and proved to have regular solutions by the ideas of Chapter 9. The notion of *inverse operator* is useful, even though it is not in general a linear operator. We define $\Omega^{-1}[X]$ as the largest Y with $\Omega[Y] \leqslant X$. Then if Ω is a linear operator whose dual is a regulator, Ω^{-1} preserves regularity, since $\Omega^{-1}[X]$ is then $-^o\Omega[-X]$. In these cases, and in many others which can be handled in a theoretical way by means of factors, it is often easy to prove that all solutions are regular, but very difficult to give regular expressions for them.

The normalizer problem of Chapter 6 is related to a class of equations about which we shall soon make a general conjecture. To solve such an equation in any particular case we use iterative techniques. Thus taking $F_0 = I^*$ we define F_1 as the maximal $F \leqslant F_0$ with $F_1 E \leqslant EF_0$, then F_2 as the maximal $F \leqslant F_1$ with $EF_2 \leqslant F_1 E$, and so on, when $\mathcal{N}(E)$ is the intersection of all the F_i. I have never found any case in which a regular expression for $\mathcal{N}(E)$ was not easily deduced from this.

Exercises

11 Find the starth root of $(xy^* + yx)^* \cap (y^*x + xy)^*$.

12 If E, F, G, H are regular, prove that the largest X with

$$EX + F \cap \frac{\partial X}{\partial G} \leqslant H$$

is also regular.

13 The *internal nth root* $\sqrt[n]{E}$ is the set of all w with $w^n \in E$. Prove that if E is regular, then $\sqrt[n]{E}$ is regular for all n, and that as n varies we find only finitely many distinct events $\sqrt[n]{E}$.

Some notes on operators

The proofs of most of our operator theorems are of such a kind that in practice they lead to very complicated algorithms, and so are perhaps best regarded merely as existence proofs – we have already made some remarks to this effect in our discussion of operators of the form $f(E^l, F^\cap, {}^\delta G)$, which suggest that the reason might be that the answers are really about as complicated as the proofs suggest. So in this section we consider only calculations of a purely theoretical kind.

The theory of factors is probably the most useful tool for proving that given operators are regulators, and the following ideas are useful. From any finite set of regular events we obtain only finitely many further events (all regular) by closing under left and right word derivation and all Boolean operations. The events so obtained we call the *associates* of the original events. The event derivates and factors of a regular event are associates of that event, and all maximal solutions of a system of inequalities $f(X, Y, \ldots) \leqslant E_i$ are associates of the E_i. The values of many operators can be shown to be associates of their arguments, and the operators thence proved to be regulators.

We can correspondingly often prove that certain operators preserve context-free languages by appealing to the factor theorem for languages (that the operator L^{lr} is a regular function of operators X^l, X^r, Y^l, Y^r, $\ldots$ where $X, Y, \ldots$ are the companions of L), although in many cases we can simply quote the result that biregulators preserve languages.

Exercises

14 Which of the operators
$$\text{Perm}[X] = \{w \mid \text{some permutation of } w \text{ is in } X\}$$

Perm'$[X] = \{w|$all permutations of w are in $X\}$
Prime$[X] = \{w|w^p \in X$ for some prime $p\}$
Square'$[X] = \{w|w^n \in X$ for every square $n\}$
are linear, and which preserve regularity?

15 Which of the operators Perm, Perm', $\bigcirc$, $\leftrightarrow$ preserve languages?

Calculations with context-free languages

Here is a simple notation for some languages. We define

$$(a, b, c, \ldots |x|^{m, n, p, \ldots} \alpha, \beta, \gamma, \ldots)$$

as the minimal solution for X of $X = x + aX^m\alpha + bX^n\beta + \ldots$, and call X a *bracket language*. Thus $(a|b)$ the solution of $X = 1 + aXb$, is $\sum a^n b^n$, and $(a, b, c, \ldots |*a, b, c, \ldots)$ is the set of all palindromic words of even length. $(a, b, \ldots |\alpha, \beta, \ldots)^*$ the solution of $D = (aD\alpha + bD\beta + \ldots)^*$ is the *Dyck bracket language*. With (,), [,], {,}, $\ldots$ for $a, \alpha, b, \beta, \ldots$ D is the set of all properly matched strings of brackets – thus $\{[][()]()\}[]$ belongs to D.

In the exercise, 'solve' means 'express the minimal solution in terms of the bracket notation'.

16 Solve $L = aL*b$.

17 Solve $X = aX^2 + bX + c$

18 Show that the L of Exercise 14 satisfies $L + ba = (a^2|ab + ba|*b^2)$, and deduce that $a(a|*b)*b + ba = (a^2|ab + ba|*b^2)$.

19 Show that the minimal solutions of

$$X = (aX + Xb)^*c \qquad \text{and} \qquad Y = Y^3 + aY^2 + bY + c$$

are $X = (a, 1|*c, cb)^*c$ and $Y = (a + c|b|*c)^*c$.

20 Show that the set of all words in a, b with twice as many bs as as, is a language.

A theorem on bracket languages

The previous exercises are instances of the theorem that every context-free language is a biregular image of a Dyck bracket language, so that we can write $L = [(a, b, \ldots |*\alpha, \beta, \ldots)^*]$. The proof we give, though a fairly long one, is itself a good example of how we can manipulate biregulators to prove general theorems. We introduce new transient letters to find a grammar for L in which each production has the form

$X \to x$ or $X \to YZ$ (x terminal, X, Y, Z transient), so our equation system has the form $X_j = x_j + \sum X_i \alpha_{ijk} X_k$, where each x_j is a sum of terminal letters, and each α_{ijk} is 0 or 1. Comparing this with the system

$$Y_j = y_j + \sum Ab_j C, \qquad A = \sum Y_i a_i, \qquad C = \sum c_k Y_k,$$

where the a_i, b_j, c_k, y_j are distinct new letters, we see that X_j is recovered from Y_j by the biregulator

$$(\sum [x_j, y_j] + \sum [\alpha_{ijk}, a_i b_j c_k])^{**}$$

and Y_i from A by right differentiation by a_i, so it suffices to tackle A.

But the above equations imply

$$A = \sum y_j a_j + A \sum b_j C a_j = a + AB,$$

say, and

$$C = \sum c_j y_j + \sum c_j Ab_j . C = c + DC,$$

say, giving $A = aB^*$, $C = D^* c$, and yielding the equations

$$B^* = (\sum \alpha_j D^* \beta_j)^*, \qquad D^* = (\sum \gamma_j B^* \delta_j)^* \qquad (1)$$

where $\alpha_j = b_j$, $\beta_j = ca_j$, $\gamma_j = c_j a$, $\delta_j = b_j$. Since A is obtained from B^* by differentiating by $a = \sum y_j a_j$ it suffices to express B^* as the biregular image of a Dyck bracket language. But defining Z by

$$Z = (\sum a_j Z \beta_j + \sum \gamma_j Z \delta_j)^*,$$

and now supposing, as we may be applying a further biregulator, that α_j, β_j, γ_j, δ_j are distinct new letters, we see that B^* defined by (1) is

$$B^* = Z \cap (\sum (\alpha_j + \delta_j) . \sum (\beta_j + \gamma_j))^*$$

and this proves the theorem, which provides the best we can reasonably expect for a canonical form for the general context-free language.

Infinite machines for context-free languages

To illuminate the structure of a simple context-free language we can differentiate by Theorem 6 of Chapter 10 to get an 'infinite machine'. Thus for $D = (aDb)^*$ and $X = (aX + Xb)^* c$ we get the machines of Figs. 14.1 and 14.2. The first provides the standard rule for well-matched brackets – 'count +1 for an opening bracket and −1 for a closing bracket, and the string is well-matched provided the count is never negative and ends at zero'. Figure 14b, after a little thought, will provide us with a slightly more complicated rule for X. In simplifying the derivatives we have used the equations $X(bX)^* = (Xb)^* X = X$. It can be proved that events represented by infinite 'ultimately periodic' machines are languages, but some care is needed in construing a converse.

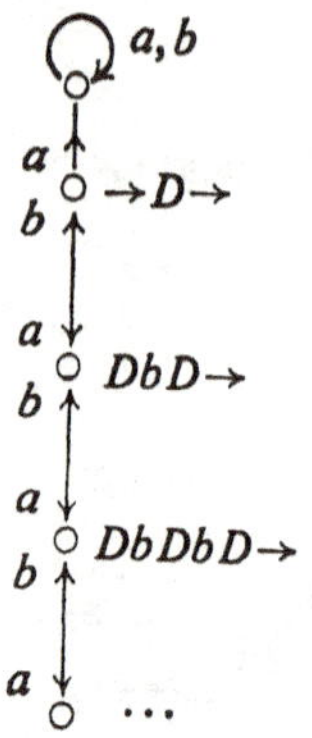

Fig. 14.1. 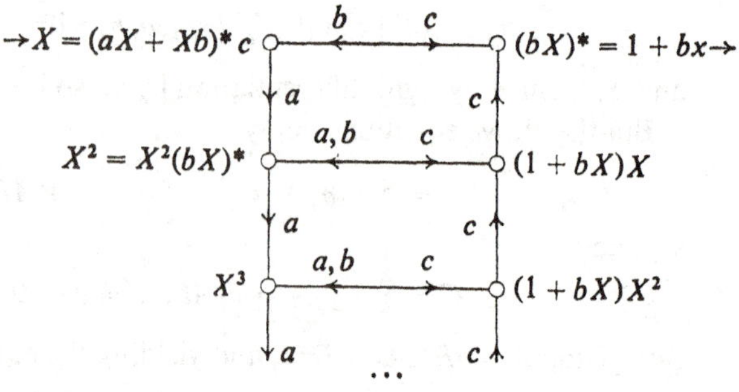   Fig. 14.2.

There is no algorithm which, supposing a language is regular, produces a regular expression for it, but if we are told that all the events appearing in the minimal solution of a set of regular equations are regular, we can find expressions for them by building 'infinite' machines (or one large partially disconnected machine) for them, and trying, as we build, all possible identifications between their states. The process will end with a finite machine or machines for which the corresponding regular expressions can be proved to be the required solutions. If the parameters in a system of regular equations are mutually commuting events, then Parikh's theorem (Chapter 11) shows that its solutions are regular functions of those events, and the above technique is often the quickest way of finding out what those regular functions are.

Exercises

21 Find infinite machines for the languages of Exercises 14–17.

22 Solve $X = X^9 + X^5 + aX + b$.

A conjecture

A system of inequalities of the type

$$\Omega[X] + \Psi[Y] + \ldots \leqslant f(X, Y, \ldots, A, B, \ldots)$$

in which Ω, Ψ, . . . are (linear) operators and f is an **S**-function, will be called a system of *semilinear inequalities*. Any such system has a unique *maximal solution* (the sum of *all* solutions). The normalizer problem reduces to such a system, namely $EX \leqslant XE$, $XE \leqslant EX$, but solutions of more general semilinear inequalities need not be regular, since for instance $X \leqslant 1 + aXb$ has maximal solution $X = (a|b)$. I conjecture that the maximal solutions of systems of the form

$$A_1 XB_1 + A_2 XB_2 + \ldots + C_1 YD_1 + C_2 YD_2 + \ldots$$

$$\ldots \leqslant f(X, Y, \ldots, a, b, \ldots),$$

in which f is regular and A_i, B_i, C_i, D_i, . . . are regular events in the letters a, b, c, . . ., are context-free languages. I suspect that the conjecture will still hold with arbitrary biregulators for Ω, Ψ, . . ., but have not tested this extension very stringently.

Exercises

23 Show that every context-free language arises as a solution of a semilinear system of the kind covered by the conjecture.

24 Prove or disprove the conjecture.

CHAPTER FIFTEEN

Some logical problems

In this chapter we sketch some of the interconnections between regular algebra and the theory of computability and recursive undecidability. A lot of the details are necessarily omitted.

Turing machines

A *Turing machine* is a finite set of things called *states*, a finite alphabet *I* of *input letters* or *symbols*, and a *transition function* f defined on some subset of $S \times I$ and taking values in $S \times I \times \{L, R\}$, where $\{L, R\}$ is the set of *directions*. The appropriate mental picture is this. We have an infinite tape ruled into squares, in each of which is written just one symbol of *I*, and in all but finitely many of the squares this is to be the particular symbol *blank*, written $\varnothing$. Above one square, containing the symbol *a*, say, is the *head* of the machine. If then the head of the machine is in state α, and $f(\alpha, a) = (\beta, b, D)$, the machine will

> *change* into state β
>
> *write b* in place of *a*

and *move* one place in the direction D (*L* for left, *R* for right).

If $f(\alpha, a)$ is undefined, the machine *stops* in state α on the symbol *a*.

We shall indicate the state of the machine and its position on the tape by writing a composite symbol

$$\ldots abcc \binom{\alpha}{a} bbaca \ldots,$$

which we call the *tape state*, and we shall omit the infinite sequences of blanks to the left and right of this expression.

Church's thesis asserts that any well-defined computational process can be performed by a Turing machine. This is not capable of formal proof, since it relates an informal concept to a formal one, but the evidence in its favour is overwhelming, and it amounts to the assertion that any such process can be programmed for a modern digital computer, supposed equipped with a potentially infinite reserve store. We cannot present all the evidence, but offer a modified form of Turing's own intuitive argument, which we find very convincing.

Let us watch a mathematician at work on a computation. We equip him with the finest of writing materials, and a very fat notebook with initially blank pages except for some near the middle containing initial information and perhaps some notes on the algorithm. Each time he turns to a new opening, he will perhaps make some small calculations, and consequently insert some new information at that opening, or rub out some information no longer required, before turning one page either forward or backward. Whatever the details, it is clear that the information remaining as a result of this process can only depend on the information previously there and the state of mind of our mathematician as he beheld it. These also determine whether he leaves by turning forward or backward, and his state of mind while so doing.

If we suppose that a page of the notebook can contain only finitely much information, and that our mathematician can be in one of only finitely many distinct states of mind, we see him in his true colours as a Turing machine! Obviously we could allow him to turn over up to 100 pages at once, say, without overstraining him, but we could not give him the freedom to turn over n pages for arbitrary n, since some values of n will be too big for his finite mind to contain! For a similar reason we could not suppose the pages to be serially numbered, since some of the numbers would be too big to fit on a page – we did say that it was to be a very fat notebook!

The insolubility of the halting problem

We sketch briefly, using Church's thesis, a proof that there is no algorithm which, given an arbitrary Turing machine M, and a tape for that machine, decides in a finite time whether or not the machine will ever stop when started at the left end of that tape in a standard initial state α. For reasons which will be apparent later, we can confine ourselves to machines with at most 100 symbols, say. If this *halting problem* were algorithmically soluble, there would exist a machine H which when started in state α at the left end of a tape consisting of (i) a coded description of M, followed by (ii) a tape t for M determines whether or not M would stop when started in state α at the left end of t, H stopping in state β if M *would* stop in these circumstances, and stopping in state γ if *not*.

But from H we could readily construct a machine H' which when started in state α at the left end of a tape consisting of just a coded description for a machine M, stops if and only if M would not stop when similarly started on this tape. The first action of H' is to duplicate its starting tape, after which it behaves like H, but where H would stop in state β, H' instead changes into some state in which it moves continually rightwards (say).

Since for every machine M, H' stops when started on the description of M if and only if M would not, we get a contradiction if we start H' on the description of H'. The reader will more readily appreciate the practicability of our constructions when he has followed some of the later arguments.

We can modify the argument to show that for any two tapes t and u there is no algorithm which, for an arbitrary machine M', decides in a finite time whether M', started in state α at the left end of t, will ever produce the tape u. For, given an arbitrary machine M, and a tape v for M, we can construct a machine M' which, when started in state α at the left end of t, first deletes t, then writes v, imitates M, and if M ever stops, instead clears the tape and sets up the desired tape u.

Normal systems

A *normal system* is a finite set of *productions* $u \to v$ in which u and v are words over a finite alphabet I, and the production $u \to v$ replaces any word of the form ut by tv, being *inapplicable* to words not of the form ut. We can imitate a Turing machine by a normal system as follows. We suppose throughout that the machine we imitate is of such a kind that in its tapes a blank symbol ($\varnothing$) is never written between two non-blank symbols. We can arrange this by letting the machine use a substitute symbol for any internal blank.

The *symbols* of the imitating normal system are the tape symbols of M together with symbols of the form $\binom{\alpha}{a}$ in which a is a symbol and α a state of M. The *productions* are those of the form

$$c\binom{\alpha}{a} \to \binom{\beta}{c}b \quad \text{whenever } f(\alpha, a) = (\beta, b, L)$$

$$\binom{\alpha}{a}c \to b\binom{\beta}{c} \quad \text{whenever } f(\alpha, a) = (\beta, b, R)$$

$$\varnothing\,\varnothing \to \varnothing\,\varnothing\,\varnothing\,\varnothing$$

$$xy \to xy \quad \text{unless } xy \text{ is the left of one of the productions above,}$$
$$\text{or } x \text{ or } y \text{ has the form } \binom{\alpha}{a} \text{ and } f(\alpha, a) \text{ is undefined.}$$

The *word* corresponding to any tape state $\ldots \varnothing\,\varnothing w\varnothing\,\varnothing \ldots$ is any cyclic permutation of $\varnothing^n w$ in which $n \geqslant 3$ and the total length of $\varnothing^n w$ is odd.

It should be clear how this normal system imitates M – it acts by peeling off two letters from the start of the word and adding them at the end, making the changes corresponding to the action of M when it can. For every complete cycle of this kind two blanks are added, so that there is no danger of the tape overlapping itself. Since the word length remains odd, the system 'changes phase' every cycle, and so M moves at least once in every two cycles. Since the normal system stops only when M does, we conclude:

Theorem 1. *There is no algorithm which determines whether or not an arbitrary normal system will ever stop when started on a given word.*

Note. The normal system here can be taken over an alphabet of only two letters a and b, for if the alphabet had at most 2^k letters we could replace these by sequences of exactly k a's and b's.

The Post correspondence problem

In the previous section the normal system imitating a Turing machine M was extravagant in its use of blanks, so that its words got longer and longer whatever the action of M. We now need a system which mimics M more closely, keeping only a bounded number of extra blanks in the word.

The new system has productions

$$ab \rightarrow cd$$
$$\varnothing \varnothing \varnothing \varnothing \varnothing \varnothing \rightarrow \varnothing \varnothing \varnothing \varnothing$$
$$\varnothing \varnothing \varnothing \varnothing ab \rightarrow \varnothing \varnothing \varnothing \varnothing cd$$
$$\varnothing \varnothing abef \rightarrow \varnothing \varnothing \varnothing \varnothing cdgh$$

whenever $ab \rightarrow cd$ and $ef \rightarrow gh$ are productions other than $\varnothing \varnothing \rightarrow \varnothing \varnothing \varnothing \varnothing$ of the old system. It is easy to see that if in the new system we start with between four and thirteen blanks in the word (still of odd total length), it retains this property after every cycle.

Theorem 2. *There is no algorithm which decides, given a normal system and two words v and w, whether or not system, if started at v, will ever result in w.*

Proof. We consider the normal system we have just constructed, imitating a machine which is to start on the tape t and end, problematically, with the tape u. We let $v_1, \ldots, v_n$ be the (finitely many) words of the normal system which represent the tape t, and $w_1, \ldots, w_n$ those representing u. Then if there were an algorithm of the type required, we could decide for each pair v_i, w_j, whether or not the corresponding normal system, started at v_i, would result in w_j, and this would imply the solution of the similar machine problem, which we have proved impossible.

Theorem 3. *The following problem (Post's correspondence problem) is algorithmically insoluble. Given a finite number of ordered pairs $(u_1, v_1), \ldots, (u_n, v_n)$ of words over a finite alphabet I of two or more letters, to decide whether or not there exists a number $k \geqslant 1$ and indices $i_1, i_2, \ldots, i_k$ such that*

$$u_{i_1} u_{i_2} \ldots u_{i_k} = v_{i_1} v_{i_2} \ldots v_{i_k}.$$

Proof. We consider the normal systems we have just constructed, and ask the algorithmically insoluble problem 'does this system, if started at w_0, lead to the word w?'. If so, there is a chain of indices $i, j, k, \ldots, q$ such that

$$w_0 = u_i t_i, \qquad t_1 v_i = u_j t_2, \qquad t_2 v_j = u_k t_3, \ldots, \qquad t_n v_k = w,$$

so that

$$w_0 \, v_i \, v_j \, v_k \ldots v_q = u_i \, u_j \, u_k \ldots u_q \, w.$$

We now add two symbols X and Y, and write the above equation in the form

$$X . u_i . u_j . u_k \ldots u_q . w Y = X w_0 . v_i . v_j . v_k \ldots v_q . \, Y$$

from which it becomes obvious that the correspondence problem for the pairs

$$(X, X w_0), (u_1, v_1), \ldots, (u_m, v_m), (w Y, Y)$$

is soluble if and only if the normal system $u_1 \to v_1, \ldots, u_m \to v_m$, started at w_0, yields w, and this we know we cannot determine. The argument uses the fact that for the normal systems we have constructed, if $w_0 v_i v_j v_k \ldots v_q = u_i u_j u_k \ldots u_q w$, then the product of the first $r + 1$ terms on the left hand side is at least as long as the product of the first r terms on the right. We can now reduce the alphabet to two letters by the same sort of coding that we have already used.

The Post Correspondence Problem provides most of the direct applications of this theory to regular algebra.

A universal Turing machine

In Fig. 15.1 we indicate a Turing machine which is capable of imitating any normal system on the letters $a, b, \ldots$, and so capable of imitating any Turing machine whatsoever. The states of this machine are at the left of the table, and tape symbols at the top. The entry gives the new state and symbol, but either of these is omitted if it happens to coincide with the old. The direction L or R is always omitted, since we have arranged that it is the same as the symbol L or R appearing in the name of the new state. Thus a completely blank entry does not correspond to an undefined transition, but rather to a situation in which the machine moves left or right without changing either state or symbol. The purpose of the four boxed entries will be explained later. We have illustrated a case with only three letters a, b, c, but the construction is quite general.

In use, the machine acts on tapes of the form

$$\ldots \varnothing \; \varnothing \; \varnothing Z Y \ldots C B A 00000 \ldots 000 abc \ldots xyz \varnothing \; \varnothing \; \varnothing \ldots$$

in which $A, B, C, \ldots, Z$ are chosen from $a^+, a^-, b^+, b^-, \ldots$, and $a, b, c, \ldots x, y, z$ are chosen from $a, b, c, \ldots$. It repeatedly emits 'signals' from the left hand part of the tape, which is a coded description of the normal system being simulated, and these signals are used to modify the right hand part of the tape, which contains the word being acted upon. In view of this behaviour, we can easily sketch the

	∅	a^+	b^+	c^+	a^-	b^-	c^-	0	−	+	a' b' c'	a	b	c
a_L^+	a_R^+,a^+													
b_L^+	b_R^+,b^+													
c_L^+	c_R^+,c^+													
a_L^-	a_R^-,a^-													
b_L^-	b_R^-,b^-													
c_L^-	c_R^-,c^-													
L		$a_L^+,0$	$b_L^+,0$	$c_L^+,0$	$a_L^-,0$	$b_L^-,0$	$c_L^-,0$							
L'								$L,-$			a b c			
a_R^+	L,a								$L,+$	$L,+$	0 0 0			
b_R^+	L,b								$L,+$	$L,+$	0 0 0			
c_R^+	L,c								$L,+$	$L,+$	0 0 0			
a_R^-								$L,-$	0			L,a'	L',a	L',a
b_R^-								$L,-$	0			L',b	L,b'	L',b
c_R^-								$L,-$	0			L',c	L',c	L,c'

Fig. 15.1. A universal Turing machine.

action of the machine by studying the left- and right-hand parts of the tape individually, noting their response in a number of situations, as follows:

In line (1) the machine is shown approaching the left of the tape in state L. It picks up the rightmost symbol A and replaces it at the extreme left of the tape, and then carries it as a signal A_R to the right end of the tape. So the tape shown will repeatedly emit the signals

$$\ldots \varnothing\, ZYX \ldots CBA000 \ldots \leftarrow [L] \qquad \text{becomes}$$
$$\ldots AZYX \ldots CB0000 \ldots [A_R] \rightarrow \qquad (1)$$

$$[q_R^-] \rightarrow \ldots 00 \oplus a'b' \ldots p'qr \ldots \qquad \text{becomes}$$
$$\leftarrow [L] \ldots 000a'b' \ldots p'q'r \ldots \qquad (2)$$

$$[k_R^-] \to \ldots 00 \oplus a'b'\ldots p'qr\ldots \qquad \text{becomes}$$
$$\leftarrow [L]\ldots 00 - ab\ldots pqr\ldots (k \neq q) \qquad (3)$$

$$[k_R^-] \to \ldots 00 - abc\ldots \qquad \text{becomes}$$
$$\leftarrow [L]\ldots 00 - abc\ldots \qquad (4)$$

$$[k_R^+] \to \ldots 00 + abc\ldots \qquad \text{becomes}$$
$$\leftarrow [L]\ldots 00 + abc\ldots \qquad (5)$$

$$[k_R^+] \to \ldots 000a'b'\ldots q'rs\ldots z \oslash \oslash \oslash \ldots \quad \text{becomes}$$
$$\leftarrow [L]\ldots 00000\ldots 0rs\ldots zk \oslash \oslash \ldots \qquad (6)$$

$A_R, B_R, C_R, \ldots, Z_R, A_R, \ldots$ in cyclic order. Here if we wish, say, $abc \to defg$ to be a production of the normal system, we arrange for the sequence of signals $a_R^-, b_R^-, c_R^-, d_R^+, e_R^+, f_R^+, g_R^+$ to be part of this cycle. Signals of the form k_R^- are interpreted as *read* signals, and those of the form k_R^+ as *write* signals. Lines 2 and 3 show how the read signals are successively compared with the first digits of the tape at the right — if there is agreement (line (2)) a prime is added to the appropriate letter, but disagreement (line (3)) causes all primes to be removed and a mark — to be deposited at the start of the tape, which causes future read signals to be ignored (line (4)). In lines 2 and 3 $\oplus$ denotes a symbol which can be either 0 or $+$, while in line (5) $\pm$ is a symbol either $+$ or $-$. If there has been disagreement at some point in the read phase just described, the first write signal changes the mark — to $+$, but otherwise it and subsequent write signals are ignored (line (5)). But when the read phase has been completely successful any write signal deletes the primed letters, which are no longer required, and deposits its message at the extreme right end of the tape (line (6)). So to imitate the normal system with productions $abc \to defg, \ldots,$ $vw \to xyz$, say, the symbols $a^-, b^-, c^-, d^+, e^+, f^+, g^+, \ldots, v^-, w^-, x^+, y^+,$ z^+ are written in reverse order at the left end of the tape, and the word on which the system acts is written in the normal order at the right end.

The machine so described will never stop, since if no production is applicable it continues a vain search indefinitely. To cure this, we introduce a new letter c, and add productions $w \to c$ for all the shortest words w which are not initial segments of words u in productions $u \to v$. Then on any long enough word there will always be an applicable production. If we now delete the boxed entries of Fig. 15.1, so that the machine stops whenever it tries to write c, or read a blank symbol, it is easy to see that in fact it stops whenever the normal system (here on two letters a, b) would.

The universality of this machine has several advantages. Since normal systems with two letters suffice for our undecidability questions, it shows for instance in the halting problem that we need consider only machines with, at most, sixteen symbols. It also shows that there is a fixed normal system for which the problem of deciding whether a word u has an infinite history is algorithmically undecidable. It should also suffice to show the reader that the sort of manipulations we used in our discussion of the halting problem can really be performed by a Turing machine.

Applications to regular algebra

Theorem 4. *There is no algorithm which, given words u, v and a biregulator Ω, decides whether or not $v \in \Omega^*[u]$.*

Proof. If we take for Ω an operator of the form $\sum {}^e u . v^r$ this would yield a solution of the similar problem for normal systems.

Theorem 5. *There exists no algorithm enabling us to decide whether two biregulators intersect, regarded as sets of ordered pairs $[u,v]$.*

Proof. The Post correspondence problem can be stated in the form 'do the operators

$$\Omega = ([a,a] + [b,b] + \ldots)^{1+**}$$

and

$$\Psi = ([u_1, v_1] + \ldots + [u_n, v_n])^{**}$$

intersect?'

Theorem 6. *There exists no algorithm enabling us to decide whether or not two biregulators coincide as operators.*

Proof. The complement of the above operator Ω, for convenience over an alphabet of just two letters a, b, can be written as a biregulator:

$$1 + \Omega : ([a,b] + [b,a]) : ([1, a+b] + [a+b, 1])^{**}$$
$$+ \Omega : ([1, a+b]^{**} + [a+b, 1]^{**})$$

The problem above can now be stated in the form 'Is this operator equal to its sum with Ψ ?'

We can also state this as a theorem on abstract **S**-algebras.

Theorem 7. *There is a semigroup with soluble word problem for which the decision problem for equality of regular expressions in the corresponding **S**-algebra is algorithmically insoluble.*

Proof. The algebra of biregulators on two letters is abstractly isomorphic with the **R**-algebra generated by symbols $A = [a,1]$, $B = [b,1]$, $C = [1,a]$, $D = [1,b]$ which satisfy the defining relations $AC = CA$, $BC = CB$, $AD = DA$, $BD = DB$. The semigroup (with unit) generated by these symbols has a trivially soluble word problem.

Theorems 6 and 7 and their proofs, are due to M. S. Paterson.

Fischer and Rosenberg have investigated the classes η_n of events accepted by n-tape non-deterministic finite automata in the sense of Rabin and Scott, and the corresponding decision problems. These are the sets of n-tupes $[w_1,\ldots,w_n]$ of words which belong to $[,\ldots,]$ in an obvious extension of our notation – that is, they are the sets of n-tupes obtainable from $[a,1,\ldots,1]$, $[1,a,\ldots,1]$, etc., by applying the obvious n-regular operations. According to Fischer and Rosenberg the following problems are algorithmically insoluble:

(1) the disjointness problem;
(2) the containment problem;
(3) the universe problem;
(4) the cofiniteness problem;
(5) the equivalence problem.

The proofs can in every case be reduced to the Post correspondence problem.

We have not space to discuss the decision problems associated with context-free languages. For an excellent treatment, see the works of Chomsky and Ginsburg.

At various times in this book we have laid emphasis on the fact that certain problems in regular algebra were algorithmically soluble. Perhaps it is the insolubility of problems immediately outside this range which accounts for the unexpected difficulty of other parts of the theory.

Solutions to some of the exercises

1 $(xy^* + yx)^* = ((xy^*)^{1+*} + yx)^* = (x(x + y)^* + yx)^*$
     $= (yx)^*(x(x + y)^*(yx)^*)^* = (yx)^*(x(x + y)^*)^* = $ desideratum.

2 $(1 + x^*y)(x + yx^*y)^* = (1 + x^*y)(x^*yx^*y)^*x^* = (x^*y)^*x^* = $
                                                   $(x + y)^*.$

We omit solutions to 3, 4 and 5 which are increasingly difficult. A proof of 3 (not from $C1$–14) is implicit in later exercises.

6   Set $E = (y^2x + xy^*xy)^*$, $F = (x^2 + yx^*)^*$, $I = (x + y)^*$, and note that $F \geqslant yI$, whence $x^*F = I$. Using these we produce the machine in the form of a table of derivates:

X	$X \geqslant 1$?	$\partial X/\partial x$	$\partial X/\partial y$
EF	yes	$y^*xyEF + xF = H$	$yxEF + x^*F = I$
H	no	$yEF + F = F$	$y^*xyEF = J$
F	yes	xF	I
J	no	yEF	J

We have omitted rows corresponding to the trivial cases I, O, xF, yEF, so that the machine has 8 states EF, H, F, J, I, O, xF, yEF. It is in fact reduced (so that *my* expertise is 100%!).

8   Call the expression $E \backslash F$, and differentiate:

$E \backslash F$	no	$y^*E \backslash (1 + y)F = I \backslash G$	$xE \backslash y^*xF = x(E \backslash F)$
$I \backslash G$	no	$I \backslash G$	$I \backslash (1 + y^*x)F = I \backslash F$
$I \backslash F$	no	$I \backslash G$	$I \backslash y^*xF = I \backslash H$
$I \backslash H$	yes	$I \backslash F$	$I \backslash H$

140

Here we have used $y^*E = I = (x + y)^*$. From the corresponding machine we read off

$$E \backslash F = (yx)^* x[x + y(yy^* x)^* x]^* y(yy^* x)^* yy^*$$

which we can simplify to $(yx)^* xx^* y[yy^* x + xx^* y]^* yy^*$.

9   As above:

$E \cap F$ yes $I \cap (1 + y)F = (1 + y)F$ $xE \cap y^* xF = x(E \cap F)$.
So if we are satisfied with any expression for $E \cap F$, we can write down $(yx)^* [1 + (1 + y)F)]$. The given expression is found by continuing to differentiate.

11 $x + xy + yx + xy^2(y + xy)^* yx$.

12   With $\Omega = E^I + F^\cap . {}^0G$, we have $X = \Omega^{-1}[H]$, $Y = \Omega^*[H]$, which are regular since Ω and Ω^* are biregulators.

13 $\sqrt[2]{E} = \sum L_i \cap R_i$, $\sqrt[3]{E} = \sum L_i \cap E_{ij} \cap R_j$, the sums over all 2-term and 3-term factorizations respectively. In general the internal nth roots are Boolean functions of the factors of E, from which the statements follow.

14   Perm and Prime are the linear operators. Perm$[(xy)^*]$ is irregular, and so Perm and Perm$'$ do not preserve regularity, since Perm$' =$ Perm^{-1}. But despite their appearance, Prime and Square$'$ do preserve regularity, since their values are respectively sums and intersections of certain internal nth roots of E, and so are associates of E.

15   Perm$[(abc)^*] \cap a^* b^* c^* = \sum a^n b^n c^n$, from which we can deduce that Perm and Perm$'$ fail to preserve languages. If L is a language and $L^{lr} = f(X^l, X^r, \ldots)$, then $\bigcirc[L] = f^\leftarrow(X^r, X, Y^r, Y \ldots)[1]$, where $f^\leftarrow$ is the reverse function of F, and so $\bigcirc$ preserve languages. It is trivial that $\leftarrow$ preserves languages.

[If L is a language in just two letters, then Perm$[L]$ can often be shown to be a language as follows. By Parikh's theorem we have Perm$[L] =$ Perm$[L']$, where L' is a sum of terms of the form $(a^p b^q + a^r b^s)^* a^t b^u$, and Perm$[L']$ is proved to be a language by rather complicated arguments like those of Exercise 20. Redko has shown that for Perm$[E]$ to be regular it is necessary and sufficient that Perm$[E] =$ Perm$[E']$ for E' a sum of terms of the form

$$(a^p + b^q + c^r + \ldots)^* w.]$$

16 $L = (a|*b)$.

17 $X = (a|b|*c)*c$.

18 $L = ab + aLL*b = ab + a.aL*b.(aL*b)*b = ab + a^2(L + ba)*b^2$
from which we get the first result. The second is an easy consequence.

20 By drawing the infinite machine, we deduce

$$M = (aMb^2 + b^2 Ma + bMaMb)*.$$

More generally, let E be the language of all words in x, y with
as many xs as ys, and $f(g, h, i, j, k)$ the regular function

$$g^{p*} \ \$ \ h^{r*} \ \$ \ i* \ \$ \ j^u \ \$ \ k^v$$

Then
$$\text{Perm}[(a^p b^q + a^r b^s)*a^u b^v] =$$
$$= f(([a, x^{2r}], [a, x^{ps}], [b, y^{pr}], [a, 1], [b, 1]))[E]$$

22 $X = (a*b)^{4*+1}$.

23 If the functions $f_i(x_1, x_2, ..., a, b, ...)$ are regular functions in
which every word involving an x_i involves one of $a, b, ...$, then
the system $X_i = f_i(X_1, X_2, ..., a, b, ...)$ has a *unique* solution,
which is both the minimal solution of the corresponding system
with $\geqslant$ for $=$, and the maximal solution of the system with $\leqslant$ for
$=$, and it is trivial that every language can be defined by such a
system.

Index